Quantum Mechanics
and
Bayesian Machines

Quantum Mechanics
and
Bayesian Machines

George Chapline

Lawrence Livermore National Laboratory, USA

World Scientific

NEW JERSEY · LONDON · SINGAPORE · BEIJING · SHANGHAI · HONG KONG · TAIPEI · CHENNAI · TOKYO

Published by

World Scientific Publishing Co. Pte. Ltd.

5 Toh Tuck Link, Singapore 596224

USA office: 27 Warren Street, Suite 401-402, Hackensack, NJ 07601

UK office: 57 Shelton Street, Covent Garden, London WC2H 9HE

Library of Congress Cataloging-in-Publication Data

Names: Chapline, George, author.

Title: Quantum mechanics and Bayesian machines / George Chapline,
    Lawrence Livermore National Laboratory, USA.

Description: New Jersey : World Scientific Publishing Co. Pte. Ltd., [2023] |
    Includes bibliographical references and index.

Identifiers: LCCN 2022042480 | ISBN 9789813232464 (hardcover) |
    ISBN 9789813232471 (ebook for institutions) | ISBN 9789813232488 (ebook for individuals)

Subjects: LCSH: Quantum Bayesianism. | Quantum theory.

Classification: LCC QC174.17.Q29 C43 2023 | DDC 530.12--dc23/eng20230111

LC record available at https://lccn.loc.gov/2022042480

British Library Cataloguing-in-Publication Data

A catalogue record for this book is available from the British Library.

For any available supplementary material, please visit
https://www.worldscientific.com/worldscibooks/10.1142/10775#t=suppl

Desk Editors: Logeshwaran Arumugam/Steven Patt

Typeset by Stallion Press
Email: enquiries@stallionpress.com

Preface

Although digital computation has enjoyed enormous successes, there remain many problems of interest where real time solutions remain out of reach. Prominent among the problems that have largely remained beyond the state of the art for digital computation are machine learning problems such as pattern recognition or decision-making in situations where the interpretation of observational data is ambiguous. It is somewhat embarrassing in this connection the mammalian brain can often resolve ambiguities in the interpretation of sensory data in real time with a footprint that is dramatically smaller than the footprint of the large-scale computers that are typically used for automated data analysis or reinforcement learning.

A pregnant question for the future of data science is what are the mathematical and engineering principles that underlie the data analysis capabilities of the mammalian brain. Along these lines, there is at the present time broad interest in whether emulation of the data analysis capabilities of the mammalian brain would benefit from the development of quantum information processing. In this book we do not promise a definitive answer to this question, but instead will review some of the potential benefits of using quantum information processing. Our primary motivation in writing this book is to bring attention to the remarkable circumstance that in many respects quantum mechanics provides a natural framework for Bayesian analysis — which, apart from the benefits that a quantum perspective may provide for using Bayesian methods to solve practical data analysis problems, may also shed light on the remarkable capabilities of mammals to evaluate risks and elect survival strategies.

What follows is essentially an elaboration of the author's 2001 *Philosophical Magazine* piece 'Quantum mechanics as self-organized information fusion', 2005 *International Journal of Quantum Information* paper 'Quantum Mechanics and Pattern Recognition', and 2007 contribution to the Berni Alder Festschrift *Quantum Mechanics and Machine Learning*. This book updates these presentations in the sense that it refines our previous notion of quantum self-organization by emphasizing the connection between optimal control/RL and solutions of integrable nonlinear partial differential equations. Our most important result is that these equations provide a basis for describing the reward function for a wide range of problems, including examples of reinforcement learning, which are very difficult to solve with conventional computational resources. These integrable systems also underscore the paramount importance of spaces of holomorphic functions of a complex number lying on a Riemann surface for representing Bayesian learning. The emergence of Riemann surfaces as an essential ingredient underscores a profound connection between cognitive science, pure mathematics, and theoretical physics.

The evolution in our views from what was presented in our original papers has especially benefited from the insights found in the papers and books of Norbert Wiener, David MacKay, Thomas Kailath, and James Rosen. In addition, the 1967 Tokyo lectures of Michio Kuga and 1997 Oxford lectures of Graeme Segal have been very inspirational. As a result, we are now much better able to explain why quantum theory is essential to understanding the mathematical principles involved in applying Bayes's formula. For the future, we hope that these insights will lead to better methods for solving practical data analysis problems.

No knowledge will be assumed on the part of the reader with respect to the current state of the art for machine learning, quantum computing, or computer-based artificial intelligence. Instead, our focus will be on the underlying mathematical relationship between quantum mechanics and Bayesian inference. Our hope is that since our description of the mathematical foundations of Bayesian learning has its own charm, and with the help of the references we cite, students and researchers will hopefully gain some new insights into why quantum devices may be useful for data analysis and artificial intelligence.

Our presentation assumes that the reader has some familiarity with the basics of probability theory and its connection with information theory; for example, as outlined in MacKay's outstanding *Information Theory, Inference, and Learning Algorithms*. Some familiarity with feedback control at the level of Astrom and Murray's or Anderson and Moore's treatises would also be very helpful. In addition, some exposure to quantum mechanics at the level of David Saxon's undergraduate textbook, or better yet Feynman and Hibbs's *Quantum Mechanics and Path Integrals* should be regarded as a prerequisite for this book. Hermann Weyl's *Group Theory and Quantum Mechanics* provides the definitive introduction to why the mathematical theory of groups and quantum mechanics are intimately intertwined.

Previous exposure to "quantum computing" is not required. For the most part our approach to quantum Bayesian learning is quite different from the approaches to information processing that can be found in the literature on quantum computing, On the other hand, occasional perusal of Nielson and Chuang's *Quantum Computation and Quantum Information* may prove helpful. By and large we will reserve the term "machine learning" to mean the extensive use of multi-layer "deep neural networks" (DNNs) for pattern recognition and reinforcement learning. However, this book does not assume that the reader is familiar with DNNs or their applications. Instead, our intent is to focus on the relationship between Bayesian learning and quantum mechanics.

Additional background for our presentation can be found in the references cited in the text. The references cited in the text are not intended by any measure to be an exhaustive listing of all the relevant literature; but instead for the most part are simply the books and papers that the author has found to be helpful. It is assumed that the reader is already familiar with linear algebra and matrices, and elementary methods for solving differential equations. A problematic aspect of our presentation for the general reader though is its extensive dependence on sophisticated mathematical results from the theory of analytic functions of complex number variables, algebraic geometry, and the theory of groups. The reader may have to spend some time "coming up to speed" with these topics in order to fully understand our presentation. As for the theory of analytic functions of a complex number variable, it would be very helpful if the reader obtained one of the standard textbooks on the theory of functions of a complex variable from Amazon. (The author is particularly fond of Copson's 1935 textbook.) For the most part, physicists and data scientists regard the theory of functions of a complex variable as *terra incognita*. Ironically, this area of mathematics is usually not ignored in engineering schools, so engineers may find our presentation more accessible.

Throughout our presentation the acronym *GP* will mean a vector with many components which are independent, independently distributed (*iid*) Gaussian random variables. The acronyms DNN, HPC, ML, MDL, MDP, NLS, ODLRO, PDE, RL, and TSP will stand for deep neural network, high performance computing, maximum likelihood, minimum description length, Markov decision process, off-diagonal long-range order, nonlinear Schrodinger, partial differential equation, reinforcement learning, topological insulator, and the traveling salesman problem (TSP). Throughout we reserve the notation $V(x)$ to mean the Bellman cost (or value) function, while the notations $v(x)$ or $q(x)$ will be reserved to mean the potential in a Schrodinger or Dirac equation. The initials BFS, GLM, HJB, KdV, KL, RH, and TO stand for Bargmann–Fock–(Irving)Segal, Hamilton–Jacobi–Bellman, Gelfand–Levitan–Marčenko, Korteweg de-Vries, Kullbach–Leibler, Riemann–Hilbert, and Togdan–Olver. We depart form the standard terminology used in the mathematics literature in one important respect: we refer to the kinematical group for quantum mechanics as the Weyl–Heisenberg group rather than the Heisenberg group, because it was Herman Weyl inspired by the work of Pascal Jordan who discovered this group.

About the Author

George Chapline's scientific career began at age 15, when he wrote a letter to Richard Feynman regarding the incompatibility of quantum mechanics and general relativity. The author's assertion – that these two theories are incompatible because quantum mechanics is intrinsically non-local, while general relativity is local – remains to this day the definitive statement of this unsolved problem. In 2000, Chapline proposed in collaboration with Robert Laughlin and his students in 2000 an explanation of how classical space-time can emerge from quantum gravity, based on the concept of quantum self-organization. This in turn has led to new perspectives in astrophysics; e.g. the hypothesis that dark matter consists of primordial black holes. Chapline graduated with a BA in Mathematics from UCLA in 1961 and a PhD in physics from Cal Tech in 1966. He was assistant professor of physics at the University of California at Santa Cruzfro 1966–1969. He has been a staff member of the Lawrence Livermore Laboratory since 1970. In 1984, Chapline won the E. O. Lawrence award for directing the experimental team that demonstrated the world's first X-ray laser. Since 2000 Chapline's research has been primarily focused on applying quantum self-organization to Bayesian inference.

Acknowledgments

The author is especially grateful to Michael Schneider for sharing his extensive knowledge of Gaussian processes and Fredholm integral equations, and to James Rose for bringing his optimization principle for inverse scattering to his attention. The author would also like to thank Jim Barbieri for discussions regarding analog neutral computing, Gennady Berman, Jonathan Dubois, and Mathew Otten for discussions regarding quantum computing, Dongxia Qu for sharing her understanding of topological insulators, and Bradan Soper for sharing his knowledge of game theory. Finally, the author is grateful to Steve Libby for discussions regarding all aspects of quantum physics.

Contents

Chapter 1

Introduction

One of the most important challenges for modern computer science is how to emulate the remarkable cognitive capabilities of the mammalian brain; particularly in situations where there is a necessity for rapidly selecting among multiple possible interpretations for sensory data (cf. Fig. 1.1). At first sight the "machine learning" techniques that are currently used to emulate the information processing capabilities of the mammalian brain would appear to lack any coherent mathematical foundation. It has long been thought [1] that "deep" artificial neural networks (DNNs), which can be either deterministic or probabilistic, might provide a pathway to emulating the cognitive capabilities of the mammalian brain. However, it turns out that these techniques are better suited to interpolation than extrapolation. Moreover, the Herculean efforts typically required to train neural networks for complex reinforcement learning (RL) [2] or feedback control [3] problems have given pause to the assumption that deep neural networks (DNNs) are the best path forward. An alternative view is that probability theory [4–6], and Bayes's theorem [5–9], provide the best framework for automating pattern recognition and decision-making.

The point of departure for this book is that, apart from the possibility [10] that the mammalian brain can estimate Bayesian probabilities, the only other circumstance where mathematical probabilities spontaneously appear in nature are physical phenomena where quantum effects are important. Therefore, it is natural to imagine [11] that the apparent ability of the mammalian brain to construct models of Bayesian inference is somehow related to quantum theory, and this

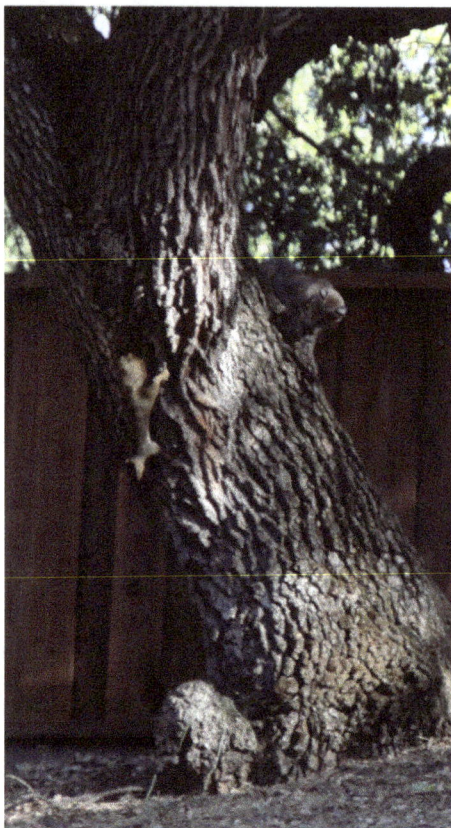

Fig. 1.1.   Natural Bayesian machines.

book can be thought of as an updated version of Ref. [11]. Our belief
in the potential importance of quantum theory for understanding
mammalian cognition is abetted by Teuvo Kohonen's discovery [13]
of the importance of self-organizing maps for understanding the organization of sensory neurons in the cerebral cortex of the mammalian
brains. However, we want to make clear from the outset that we
do not subscribe to the notion that the mammalian cerebral cortex
is a "quantum computer" in the sense that this term is commonly
used [14]. Kohonen's maps involve holomorphic analytic functions of
a complex number variable [13] that are very similar to those that
appear in analytic solutions of the 2D Schrodinger equation [15].
(Holomorphic functions are smooth functions of a complex number

whose only singularities are zeros.) Focusing on the importance of holomorphic functions is a central theme of our presentation, and is what allows us to tie together Bayes's formula, integrable dynamics, and mammalian cognition.

Putting aside the question as to how the mammalian brain actually works, we will adopt the point of view that the similarity of the space-time structure of quantum dynamics and Markov decision processes (MDPs) [16,17] is an important clue that should not be ignored. At center stage here are the observations of Aharonov *et al.* [18] regarding the relationship of past and future measurements in quantum theory. Historically, it was first observed by Schrodinger himself in the 1930s [15] that MDPs can be written in a way that combines forward and backward propagating processes. Remarkably both Bayes's formula and Todorov's extension of Bayes's formula [19] to include control processes [3,19,20] also involve forward and backward MDPs.

The foundation for our presentation is Bayes's epochal formula for conditional probabilities [5,6], which provides a rational basis for data analysis and decision- making in virtually all contexts. In poetic language, Bayes's formula can be written:

$$P(\theta|d,\alpha) = \frac{Likelihood \times Prior}{Evidence}, \tag{1.1}$$

where $P(\theta|d,\alpha)$ for the parameters θ *associated* with a particular data model α chosen from a set $\{\alpha\}$ of possible data models is correct given a set $\{d\}$ of input data and the *Likelihood* $P(d|\alpha,\theta)$ and *Prior* $P(\alpha|\theta)$. The *Prior* is the *a priori* probability that a particular explanation α is correct, and the *Evidence* is the probability of the input data summed over all possible explanations. Although Bayes put forward his formula in the 18th century, widespread appreciation of the great scientific and practical significance of Bayes's formula did not arrive until late in the 20th century. Fortunately, the theoretical importance of Bayes's formula for data analysis is now widely recognized. Indeed, it is now widely accepted [5–8] that formula (1.1) provides the basic framework for solving virtually all problems involving understanding sensory data and observation-based decision-making. In the following, we will refer to the enterprise of evaluating the conditional probabilities on the r.h.s of Eq. (1.1) for the purposes of solving these types of problems as "Bayesian inference".

Unfortunately, even though the fundamental importance of Bayes's formula is now widely appreciated, the use of Eq. (1.1) toward obtaining accurate estimates for $P(\theta|d, \alpha)$ in all circumstances of interest has remained limited. This is especially true, for example, in situations where there are hidden factors [7]. As has been emphasized by MacKay [6], one reason for lack of progress in this direction is that in general all three of the factors on the r.h.s. of Eq. (1.1) must be evaluated to obtain a complete understanding of input data. In addition, the conditional probabilities that appear in the numerator and denominator of the r.h.s of Eq. (1.1) cannot in general be evaluated in real time with state-of-the-art computational algorithms.

In the 19th century, Maxwell had emphasized the importance of probability theory for understanding physical phenomena [23], but folklore attributes to Helmholtz the notion that the human brain can estimate the probabilities for various alternative explanations for sensory observations. As it happens, Helmholtz's suggestion regarding the ability of the mammalian brain to estimate probabilities lay dormant for more than a century. In contrast with theoretical physics, where there are validated models for almost all physical phenomena of any practical importance, the mathematical principles underlying the cognitive capabilities of the mammalian brain have remained elusive. Observations of the natural behavior of mammals in the wild (cf. Fig. 1.1) do suggest that the mammalian brain can construct reinforcement learning-like strategies for dealing with adversity in the real world. One of our primary motivations for writing this book is to explain the mathematical principles underlying the observed abilities of mammals to evaluate Bayesian probabilities.

For a glimmer of hope that improved understanding of how the mammalian brain evaluates Bayes's probabilities might indeed lead to better methods for artificial data analysis, the data science community is indebted to Harvard mathematician David Mumford [24] for having put forward the plausible hypothesis that the intellectual capabilities of the mammalian brain evolved in such a way as to endow the mammalian brain with the capability to construct models for the world which are the simplest from the perspective of the amount of information needed to describe the model. Mumford's suggestion is similar to the suggestions of Kolmogorov and Chaitin [25] that the value of any approach to data analysis can be assayed by the algorithmic complexity of the computer program needed to implement the model. The phase "minimum description length" (MDL) for these ideas was introduced by Rissanen [26]. As it happens, no practical method for implementing the MDL principle with existing computational resources has yet been found; however, the Helmholtz machine described by Dayan, Hinton, Neal, and Zemel [8,9] is constructed around the conceptually elegant idea that the information cost for data analysis can also be interpreted as the free energy of a physical spin system.

A fundamental result in classical statistical physics is that a Maxwell–Boltzmann distribution for the relative occupations of the energy levels for a physical system in thermal equilibrium is a consequence of minimizing the non-equilibrium free energy of the system. Hinton *et al.* point out that Bayes's formula, Eq. (1.1), can then be interpreted as a Boltzmann-like distribution for configurations of the Helmholtz machine, where the configuration "energies" are the negative logarithms of the *Likelihood* factors in Eq. (1.1). This is analogous to the fact that in ordinary statistical mechanics this Boltzmann distribution can also be characterized as the probability distribution P_α which minimizes the free energy $\underline{F} = E{-}TS$, where E is the average of the negative log-likelihood and S is the entropy associated with P_α. The net result is that the *Evidence* factor in Eq. (1.1) can also be regarded as the non-equilibrium Helmholtz-free energy $F[P(x)]$ of a "physical" system:

$$Evidence \equiv \exp\left\{ -\sum_\alpha [P_\alpha E_\alpha - (-P_\alpha \log P_\alpha)] \right\}, \qquad (1.2)$$

where P_α is the conditional probability that appears on the l.h.s of (1.1). The first term on the r.h.s of Eq. (1.2) is the expected "reward"; i.e. the sum over paths of a negative exponential of an integral of a function $q(x)$ which encourages the controller to visit more likely states (cf. [19]). The second term on the r.h.s of Eq. (1.2) is analogous to the term that represents the contribution of entropy to the thermodynamic free energy. In the context of classical statistical mechanics [23] minimizing the free energy can sometimes be realized by minimizing the expected energy, and at other times realized by maximizing the entropy term. In a similar way in the context of Bayesian inference when there are multiple data models, the *Evidence* factor in Eq. (1.1) may in some cases be maximized by choosing a model whose expected likelihood (the 1st term in the free energy in Eq. (1.2)) is high, while in other cases one would like to maximize the entropy associated with the probability distribution P_α for various models (the 2nd term on the r.h.s of Eq. (1.2)). Hinton *et al.* also proposed a specific algorithm, the wake–sleep algorithm [8], for minimizing the Evidence factor in Eq. (1.1). In practice, this algorithm amounts to minimizing the Kullback–Leibler (KL) divergence [27], i.e. the difference between the entropies of the two Helmholtz machine networks expressed in terms of their respective probability distributions for their Ising spin degrees of freedom. The KL divergence plays a central role in minimizing the information description for the recognition and data description networks in the Helmholtz machine [8] by forcing the information cost of their respective representations for the data to be equal.

The KL divergence also plays a central role in measuring the information cost Bayesian search for an object in an unknown location [28], and in Bellman's dynamic programming formalism [20] for optimal control. A central feature of Bellman's dynamic programming formalism was introduction of a performance index, or "cost function" $V[x(t), u(t)]$ for quantifying how well the controller for feedback control or agent for RL is performing in efforts to bring a system or environment to a desired final state $x(\mathrm{T})$. Although Bellman didn't initially phrase it this way, it was eventually realized [28] that, apart from a "reward function" [30], Bellman's value function represents the rate at which, as a result of observations, information is gathered about the likely results of future observations. As noted by Tororov [30] and Kappen [31], the evolution of the Bellman function taking a small step along a random path can also be represented as the negative exponential of a "reward function" $q(x, t)$ describing the likelihood of a state x given a previous or concurrent control action u. The Bellman function $V(t)$ describing the information cost moving from a state $x(0)$ to a state $x(T)$ will then be given by a negative logarithm sum over all paths of an exponential of minus the reward function $q(x, t)$:

$$e^{-V(T)} = P \exp \left\{ -\int_{x(0)}^{x(T)} q[x'(t)]dt \right\}. \qquad (1.3)$$

The symbol P means averaging over all the possible paths going from x at $t = 0$ to $x(T)$ at $t = T$. The optimal path will be defined by the path where the integral over $q(x, t)$ is minimized. However, because of the necessity of exploring many paths in Eq. (1.3), the change in $V(t)$ along the optimal path will have an additional cost term, the "control cost", which encourages the controlled history of the system to lie near a path that might be attributed to the system dynamics without controls and can be identified as a KL divergence. In an earlier paper [19], Todorov showed that the loss rate for the optimal Bellman cost function can be written as a sum of $q(x, u)$ and the KL divergence term, which together are minimized as w.r.t $u(x)$. This in turn leads to an expression for $V(t)$ as the backward filtering probability for the state of the system given all previous observations of the system. By expressing the Bellman function $V(t)$ in terms of a probabilistic chain of actions (cf. [19]), the sum in Eq. (1.3) can also be evaluated using Monte Carlo methods [31]. However, this can be very time-consuming, if not intractable.

Taking to heart the similarity between the space-time structure of MDPs and quantum dynamics that was one of our original inspirations [16], one might guess that Eq. (1.3) can be faithfully emulated using Feynman's "sum over paths" interpretation for quantum mechanics introduced in his 1942 PhD thesis [32]. In this interpretation of Eq. (1.3), the action function appearing in the exponent of Feynman's sum over paths replaces the reward function $q(x, u)$ that appears in Todorov's formulation of optimal control [19], while the control variable $u(t)$ is represented by Dirac's momentum operator $-i\partial/\partial x$. The r.h.s of Eq. (1.3) would be then replaced by a sum over quantum paths where the real exponents, the r.h.s of Eq. (1.3), are replaced by $iq(x)$ where $i = \sqrt{-1}$. In this quantum interpretation of Eq. (1.3) the sum over real negative exponentials is replaced by Feynman's sum over quantum paths expression [32] for a quantum propagator describing the translation along a path $x(t)$ of a two-component wave function $\Psi(x)$:

$$T(x, y|\lambda) = P \exp \left\{ -i \int_x^y \tilde{q}(x', t|\lambda) dx'(t) \right\}, \qquad (1.4)$$

where the P symbol means time ordering of path segments, λ is a spectral parameter, and $\tilde{q}(x, t)$ is a 2×2 matrix (The necessity for two-component wave function is discussed in Section 5.5). The potential $\tilde{q}(x, t)$ will guide the evolution of the wave function in much the same way as was originally contemplated by Schrodinger for the phase of his wave function [33]. An optimization principle that will cause the amplitude in Eq. (1.4) to become focused on a single final state is known as Rose optimization [34]. The two components are necessary [35] if one wants to represent the fact that in contrast with classical mechanics, quantum dynamics can also allow for simultaneous forward and backward propagation.

In the sum of terms corresponding to the r.h.s of (1.4) the contribution from paths very different from the optimal path will tend to cancel one another. Thus, in the classical limit the only term in the sum that should survive is the one where the integral of the reward function for the initial to the final is minimized. Unfortunately, it is not at all straightforward to explicitly construct a classical path from a quantum path integral. Any attempt to construct a classical path as the classical limit of Feynman's path integral immediately runs into the immediate difficulty that the classical equation of motion for any system treats both the position and momentum on the same footing as classical variables, whereas in quantum mechanics these variables are treated in very different ways. This makes evaluating the classical limit of a Feynman path integral very tricky. Indeed, in the 1920s the founders of quantum mechanics were very puzzled as to why well-known results in classical mechanics could not be obtained in any obvious way as the limit of Schrodinger's wave equation [33] when Planck's constant of action $\hbar$ was assumed to approach zero. For example, the problem of how to obtain Kepler's laws for planetary motion from Schrodinger's wave mechanics wasn't solved until the late 1980s [37]. The root of this difficulty is while the position variable is treated as a classical variable in wave mechanics, the momentum variable is regarded as an operator acting on the wave function. This difference can allow a control variable represented as a derivative of the wave function to have large excursions from its value along an optimal path. This happens, for example, if the passive dynamics for the system has a classical turning point [71]. One of the main themes in our presentation will be that this difficulty can be resolved if Feynman's original path integral for the quantum dynamics of a system is modified to allow for frequent measurements of the control variables, which are typically represented as momentum variables. This is the strategy used in adaptive optics [38] and "measurement-based quantum computing" [39] to arrive at a specified final state. The bottom line is that by including a theory of measurement quantum mechanics does seem to show promise for evaluating Eq. (1.4).

Quantum theory began to take form at the end of the 19th century as a result of Max Planck's introduction [40] of the quantum of light in 1900 in connection with the problem of understanding the spectrum of thermal radiation emerging from an oven. There was already an appreciation at the time of Planck's paper that there were a variety of physical phenomena, e.g. the dependence of the chemical and spectral properties of atoms on their atomic number, the spectrum of thermal radiation, radioactivity, etc. that were refractory to explanations based on classical physics. However, Planck's focus on the problem of understanding the entropy of thermal radiation turned out to be pivotal for the future of physics. Following Planck's unveiling of light quanta, it was soon realized, largely as a result of the work of Bohr and Sommerfeld [41], that Planck's discovery had profound implications for our understanding of atomic matter. Quantum mechanics emerged in the 1920s because of a desire to extend the Bohr–Sommerfeld quantum theory, which had only been successful for simple (actually "integrable") physical systems, to all types of physical systems. As prophesied by Dirac [35], quantum mechanics does in fact appear to provide us with a mathematically consistent framework for understanding all known natural phenomena. Quantum mechanics made its debut [42] in 1925 with the two simultaneous papers of Dirac and Heisenberg, Born, and Jordan. These papers provided a foundation for theoretical physics where matrices were used to represent physical quantities. Initially the physical meaning of this "matrix mechanics" was rather obscure, although it soon became clear [43] that the epistemological flaw with classical physics lay with the implicit assumption that the variables used in classical physics, e.g. the position of a particle or the magnitude and polarization of an electric field, could — at least in principle — be simultaneously measured with arbitrary precision. Before 1925 it had always been imagined that physics should be directly based on measurable quantities. Heisenberg's great achievement [43] was his "uncertainty principle", which explained that the flaw in classical physics lay in the tension that always exists between the way experimental measurements are carried out — particularly when atomic phenomena are involved — and the desire that physics should be based entirely on physical laws that were completely independent of the way measurements are carried out. In the 1925 papers of Dirac and Born, Heisenberg, and Jordan [42] it was proposed that this tension could be resolved by using matrices which satisfied certain commutation relations rather than scalar variables to represent physical quantities such as position and momentum. These commutation relations were given a precise definition and elevated to a mathematical group by Herman Weyl [44]. This 3D group, which we will refer to as the Weyl–Heisenberg group, provides a mathematically precise definition of quantum kinematics. (In the pure mathematics literature, this group is referred to as the Heisenberg group, even though, inspired by the 1925 work of Pascal Jordan [42], this group was originally introduced by Herman Weyl. Weyl's book [44] provides a nice introduction to this group.)

As it happens though, despite Heisenberg's extraordinary insight into the *raison d'etre* for quantum mechanics, the physical meaning of quantum mechanics has remained somewhat enigmatic [45]. This veil of mystery was partially lifted in the spring of 1926, when, while trying to understand how collisions between atomic particles can be understood within the framework of the matrix mechanics of Heisenberg *et al.*, Max Born hit upon the idea [46] that the absolute square of the wave function which appears in Schrodinger's wave equation formulation of quantum mechanics [33] represents a probability density for the values of the argument of the wave function; e.g. the position or momentum of a particle. This is surely one of the most important discoveries in the entire history of science! For our purposes we will rely on the definitive description of the relationship of quantum theory and probability theory provided by Feynman and Hibbs [47].

Although the physical meaning of quantum mechanics remains somewhat mysterious, there is no doubt that quantum mechanics has led to an enormously better understanding of Shannon information. It is noteworthy in this respect that while the concept of entropy plays a central role in both practical thermodynamics and information theory, no way is known for defining what the absolute entropy of a physical system means without quantum mechanics. In engineering practice one can follow Carnot [22] and define the entropy of "working fluids" using the phenomenological specific heat of these fluids. However, if one follows Boltzmann [23] and tries to define entropy of a physical system as the number of microscopic states corresponding to a macroscopic state, then in almost all cases of interest one requires quantum mechanics to define the Boltzmann entropy.

This is true for both kitchen ovens and the universe. (The entropy of the universe is to a very good approximation just the number of cosmic microwave photons per gram of dark matter.) As was emphasized by Planck in his original work on thermal radiation [40], one of the most satisfying consequences of quantizing the energy levels of a system is that the absolute entropy of any physical system acquires a well-defined combinatorial definition. The combinatorial definition of the entropy of thermal radiation provided by Planck suggests that quantum theory could be relevant to representing the information theoretic aspects of Bayesian learning.

At first sight this may appear implausible because the equations of quantum mechanics, e.g. the Schrodinger wave equation, are by themselves deterministic, and therefore there is no obvious mechanism for representing the gathering of information required for Bayesian learning. On the other hand, there is an underlying randomness associated with the choice of quantum paths in the path integral formulation of the Schrodinger equation. In addition, including the measurement process into a quantum description of the dynamics of a system apparently does offer the possibility of introducing the randomness represented by the conditional probabilities in Bayes's formula. There have been several attempts (see e.g. [48]) to describe the effects of measurements by modifying the Schrodinger equation so that it is no longer deterministic. However, there is as yet no universal agreement as to which of these stochastic extensions of Schrodinger's equation would be the canonical best choice. For our purposes we will follow the ideas of Schwinger [49], Caldeira and Leggett [50], and Keldysh [51] regarding how to describe relaxation processes due to measurements within the framework of quantum mechanics. In particular, we will make use of the double path integral description of interacting quantum systems due to Feynman and Vernon [47] (see also Appendix D).

Kappen [31] pointed out that as a function of the level of innovation noise (the difference between observations of the state models for a system) the Bellman dynamic programming equations change from being deterministic at low noise levels to being explicitly stochastic at high noise levels. This transition is reflected in the relative contributions of the reward function and KL divergence to the Bellman function loss rate. At low noise levels, the KL divergence term can be neglected and the Bellman loss rate will be determined by a reward function that is independent of the innovation noise. On the high innovation noise side of Kappen's transition, the typical objective of optimal control and RL will be to relax the KL divergence between the observed controlled dynamics and desired passive dynamics of a system, which of course has an information theoretic flavor.

Our quantum approach to stochastic control problems will be based on the notion that the Helmholtz machine [8,9] provides a kind of Rosetta Stone for translating Bayesian inference into the language of quantum theory. The "Greek to Coptic" part of this Rosetta Stone is the identification of the controller/agent in a feedback loop as the recognition network in the Helmholtz machine, and the system/ environment as the unsupervised data model generation network of the Helmholtz machine. The wake–sleep algorithm [9] reconciles these forward and backward representations of the data in such a way that the output of the data generation network relaxes to a good approximation for the input data. At the same time the wake–sleep algorithm minimizes the innovation noise seen by either network, which can also be interpreted as minimizing the Bellman–Issacs loss function for a Markov game [53]. The "Coptic to Hieroglyphic" part of the Rosetta Stone amounts to replacing the probabilistic Ising spin degrees of freedom in the original Helmholtz machine with Θ-functions [54–57]. These multi-variable analytic functions play an important role in the theory of Riemann surfaces. Riemann surfaces are smooth 2D surfaces that can also be parameterized by a single valued complex number, and turn up in a surprising variety of contexts in pure mathematics. One way of visualizing Riemann surfaces, due to Solomon Lefshetz [54], is that Riemann surfaces can be represented as the locus of a homogeneous polynomial in a higher projective space. Mumford later realized [55,56] that these coordinates, known as "Θ-functions with rational characteristics" [54–56], provide a representation, of the Weyl–Heisenberg group [44]. This identification of Riemann surfaces as "algebraic varieties" (which is the mathematical term for any manifold that can be described as the locus of a homogeneous polynomial in a projective space of some dimension) ushers in quantum mechanics *deus ex machina*, and provides a link between the theory of Riemann surfaces, optimal control, and quantum mechanics.

Perhaps the first person to sense that there is a deep connection between feedback control and quantum mechanics was Freeman

Dyson. In 1975, Dyson noticed [58] that at low light intensities where photon noise becomes important, the feedback equations of adaptive optics are formally identical with the theory of inverse scattering for the 3D Schrodinger equation. In Dyson's approach to adaptive optics, the effect of the atmosphere on a flat 2D wave front is observed by using a phase sensor that allows for observation of arbitrary 2D correlations between the atmospheric noise in different optical channels. These equations are a 3D generalization of the equations developed in the 1950s by Gelfand, Levitan, and Marčenko [59,60] for the purpose of finding the potential of the 1D Schrodinger equation based on scattering data, (see Appendix B). Dyson's discovery of a connection between adaptive optics in the presence of photon noise and the 3D Schrodinger equation naturally stimulated interest in why so seemingly disparate topics are connected, and a full resolution of this puzzle remains to this day. Our aim though is somewhat different than just understanding scattering solutions of the 3D Schrodinger equation. As in our original paper [36], we will be focused on regression between observations and models for an entire control history which terminates in a desired history.

In the following, we do not claim to prove that translating probabilistic models for optimal control such as Bellman's dynamic programming into the language of quantum mechanics necessarily provides better results than what might be achieved with conventional computational resources. However, we do wish to emphasize some *ab initio* advantages that quantum amplitudes enjoy in comparison with conventional probabilistic representations for Markov decision chains (MDPs). One prominent advantage is the elegant way in which quantum amplitudes can capture causal relationships. This is very challenging [62] for conventional machine learning techniques; especially in cases where the computational model aspires to "artificial intelligence" [63]. As was noted by Feynman in his Nobel Prize winning paper [64] introducing a relativistic quantum theory of photons interacting with electrons and positrons, there is no natural way to combine causal and anti-causal influences within the framework of classical electrodynamics (a footnote in [64]). On the other hand, in his theory of quantum electrodynamics [64] Feynman introduced a way of combining forward and backward in time propagation for electrons that takes full advantage of the fact that the relevant quantum amplitudes can be regarded as smooth functions of the particle momenta regarded as arbitrary complex numbers. Later, it was realized [65] that, independent of any underlying Hamiltonian, general quantum amplitudes for transitions between elementary particle states can always be regarded as smooth analytic functions of the momenta of the incoming and outgoing elementary particles regarded as complex numbers. This behavior is peculiar to quantum theory and reflects a fundamental equivalence between the analytic behavior of quantum transition amplitudes as a function of momentum variables and the causal structure of operator commutators in relativistic quantum field theory [66]. Causality is only the tip of the iceberg though for the shiny gloss that analytic behavior as a function of momentum variables provides for quantum amplitudes.

During the time he was a post-doctoral fellow at the Bohr Institute, Lev Landau discovered a book in the library describing how the special analytic functions that are important for finding exact solutions of the Schrodinger equation, especially in two-dimensions, could be expressed as integrals where their extension to an analytic function of a spectral parameter regarded as a complex number was transparent. Landau's loitering in the library at the Bohr Institute led to his Appendix to *Quantum Mechanics* [15], which set the stage for the notion that is now in full bloom that the theory of functions of a complex number variable and quantum mechanics are deeply intertwined (see e.g. [54,55]). What is important for us is that the integral representations for analytic functions rediscovered by Landau could also be obtained as solutions of a Riemann–Hilbert (RH) problem [67–69] (see Appendix B for a discussion of the RH problem). This type of problem can often be solved analytically using the Cauchy integral representation for analytic functions of a complex variable [67]. As shown by Its [68], the Cauchy theorem for analytic functions also provides a path from the integral representations for special functions described in Landau's Appendix to *Quantum Mechanics* to exact solutions for certain nonlinear integrable PDEs.

A simple example of this usefulness of the special analytic functions defined by a RH problem is the modified Airy functions (MAFs) [70], which provide a good approximation to the solution of 1D Schrodinger equation for any potential — even one with a classical turning point [71]. This property is a prerequisite for being able to use a quantum path integral to define a classical trajectory. As it happens, the properties of analytic functions like those listed in Landau's Appendix [15] also seem to provide the necessary ingredients for a universal underlying structure for feedback control and RL. A hint that the special analytic functions listed by Landau might indeed be useful for describing feedback control and RL is provided by the intimate connection [72] between the KdV equation and the Kalman filter. More generally, the analytic functions delivered by the RH method provide a basis for representing both the reward function and innovation for feedback control and RL problems. Of course, this leaves open the question as to how one can realize in practice the minimization of the innovation noise; i.e. minimizing the difference between the trajectory of observed observations and the expected history for a system or environment based on a model.

The great achievement of Graeme Segal (who held the Astronomy and Geometry chair at the University of Cambridge) and George Wilson in this connection [73] was to "geometrize" the inverse scattering method [73,75] for solving the KdV equation by constructing a distinguished space for meromorphic functions defined on a subset of a Hilbert space consisting of linear sum of two vector spaces of square-integrable holomorphic functions: "Hardy spaces" (named after the Cambridge U mathematician who discovered Ramanujan). Meromorphic functions are rational functions of holomorphic functions, which are smooth functions of a complex number variable whose only singularities are isolated zeros. With their construction, Segal and Wilson introduced to data science the importance of basing computational approaches to stochastic estimation and feedback control on the construction of a special holomorphic function, known as the τ-function, defined on a canonical Riemann surface. A canonical Riemann surface is a universal feature of integrable dynamics [73,77], and the recognition of its importance for integrable systems is a momentous development for data science because the topological obstructions to conventional Monte Carlo regression [78,79] can be "untangled" if the labels for observed and model states for a system or environment can be lifted to curves on Riemann surface with sufficiently large genus [80].

Our formal presentation of how quantum mechanics might be used to emulate Bayesian learning begins in Chapter 2 with a recounting of six fundamental discoveries which provide the guideposts for our path to quantum Bayesian learning. Pride of place naturally belongs to Thomas Bayes's famous formula, Eq. (1.1), for how the implications of observations can be captured as posterior probabilities [5]. After Bayes's formula perhaps the most notable discovery was Norbert Wiener's noise filter [81], which supplanted the least squares regression method for interpolating data that originated with Gauss and Legendre [82]. The original least squares method didn't distinguish between signal and noise, and therefore didn't necessarily provide a good estimate of the underlying "signal". This deficit was addressed in Wiener's wonderful 1942 paper [81] introducing the "Wiener filter", which is the seminal spring from which essentially all the analytical methods we will discuss for data analysis flow. For example, an immediate dividend of Wiener's approach to signal analysis was the very successful Kalman–Bucy model [3,83] for feedback control. The Kalman–Bucy filter does make some simplistic assumptions, such as linear time evolution in the absence of control actions, which in principle were later removed by the development by Richard Bellman of his dynamic programming approach to optimal control [20]. Unfortunately, Bellman's approach is typically very difficult to implement in practice. In some cases, Bellman's cost function $V(t)$ can be determined by numerically solving a linear differential equation (see e.g. [21]), while in other cases, e.g. complex RL problems, $V(t)$ can only be determined using machine learning techniques such as deep neural networks. (The Bellman's cost function is also sometimes referred to as a "value function". This nomenclature is awkward in that the object of optimal control is to minimize the Bellman function.) It happens that Bellman's cost function is intimately connected with information gathering [27,28]; so minimizing the cost also means choosing actions which maximize the collection rate for information with respect to an optimal choice of control actions. For our final choice for a fundamental discovery to be singled out, Chapter 2 concludes with the observation that Feynman's quantum path integral [32] can in principle be used to solve the traveling salesman problem. This observation draws attention to the possibility of using the natural tendency quantum paths to fill up all of configuration space to resolve combinatorially difficult aspects of Bayesian model selection.

Chapter 3 is an introduction to the Mumford–Rissanen Minimum Description Length criterion [26] for selecting from a variety of possible explanations for a given set of observations, the explanation that from the point of view of information theory is the most economical. This principle is a legacy of William of Ockham, who early in the 14th century [85] put forward one of the fundamental tenets of science: that the best explanation for a physical phenomenon is usually the simplest. It is perhaps counterintuitive that a principle of physical science should underlie data analysis. However, William's principle of minimizing the complexity of the explanation for a set of observations is at the heart of the notion that Bayes's formula provides the logical basis for solving a variety of problems including stochastic estimation, Bayesian searches, and feedback control. The maximum likelihood method [6], which is widely used to solve these types of problems, short circuits the full use of the Bayes formula by looking only at ratios of the likelihood factor in Eq. (1.1). However, as has been emphasized by McKay [6], simply looking at the likelihood that a model for the data yields a particular set of observations can lead to serious errors when one must choose the model from an ensemble of *a priori* approximately equally plausible models. Reflecting Mumford's insight [23] regarding the MDL principle and mammalian cognition, McKay's "Occam razor" factor [6] is possibly the best metric yet proposed for guiding data model selection. Chapter 3 concludes with a brief account of how the search for methods for dealing with hidden factors [7,29] led to the Helmholtz machine [8], which provides a logical framework for how conditional probabilities which reflect the MDL principle might be computed as Markov decision chains.

Chapter 4 focuses on control theory [3], and in particular on the deterministic limit of Bellman optimization, known as Pontryagin control [86,87]. The Pontryagin procedure for realizing the deterministic limit of optimal control is somewhat different than the Euler–Lagrange variational procedure described in textbooks for obtaining the classical equations of motion by minimizing the Maupertuis action (see e.g. [88]). The Euler–Lagrange method for obtaining the equations of motion for classical mechanics differs from the procedure for obtaining the optimal path for feedback control by minimizing the Bellman cost function in that the Euler–Lagrange method only demands uniform convergence for the positions' classical trajectory dx/dt, whereas the Pontryagin limit of Bellman optimization demands simultaneous uniform convergence in both the system variables $x(t)$, dx/dt, and control variables $u(t)$. Understanding the relationship between Pontryagin control, classical dynamics, and the classical limit of a Feynman path integral for a quantum system will be an important part of our presentation. The bottom line is that while extracting the classical limit of a quantum path integral can be very tricky, the resultant equations can mimic the Hamiltonian dynamics of Pontryagin control. Chapter 4 also introduces "Lie–Poisson dynamics" [89]; a class of analytically solvable examples of Pontryagin control which arise from the action of a continuous group on itself regarded as a smooth manifold. Lie–Poisson dynamics has some interesting practical applications such as spacecraft control and illustrates how special analytic functions such as elliptic integrals can naturally provide solutions for control problems.

Chapter 5 is the centerpiece of our presentation. Our main focus is on the use of the inverse scattering method to solve nonlinear PDEs such as the KdV and NLS equations. Of particular interest to us is the construction of Segal and Wilson [72–73], who were among the first to connect the problem of finding exact solutions of the KdV equation with evaluating the Riemann Θ-function, which is a periodic analytic function on a Riemann surface.

In the Segal–Wilson construction, the input data and data features are represented by elements of vector spaces, known as "Hardy spaces", of square integrable analytic functions of a complex number whose domains are, respectively, the inside and outside of a closed curve on a sphere. Obviously, the simplest version of this setup is when the closed curve is the equator, and the two Hardy spaces are related by complex conjugation. The KL divergence will vanish when the probability distributions for the states constructed from the two Hardy spaces are identical. This supports our identification of the Helmholtz machine as a Rosetta stone for translating Bayesian inference into quantum language, as well as providing new insights into what mathematical principles underlie the remarkable cognitive capabilities of the mammalian brain.

One might imagine that analytic solutions of the KdV or NLS equations are too specific to encompass all feedback control or RL problems of interest. However, it was realized by David Hilbert at the beginning of the 20th century that there is a remarkable similarity between Galois's theory of solvability of algebraic equations [92] and the role of meromorphic functions, i.e. rational combinations of holomorphic functions, in finding exact solutions of integrable nonlinear PDEs. This connection with Galois theory transports us from the world of special analytic function solutions of the KdV and NLS equations to a world of profound and very general mathematical objects such as Galois groups. As was realized by Hilbert [93], the analog of the subgroups of the group of permutations of the roots of a polynomial that plays a central role in Galois's theory of solvable algebraic equations are the subgroups of the group of homeomorphisms for a Riemann surface.

Chapter 6 lists some quantum tools which seem to be pertinent for improving our understanding of Bayesian learning. The most prominent of these tools is the flexibility in choosing a representation for the Weyl–Heisenberg group [44] that is best suited for the application at hand. For the feedback control and RL problems that are our primary focus in this book, it seems that representations whose elements form a reproducing Hilbert space of holomorphic functions [94–96] will be of special interest. (Holomorphic functions are smooth analytic functions of a complex number variable defined in some domain of the complex plane whose only singularities are zeros.) Among representations of the Weyl–Heisenberg group, those involving spaces of holomorphic functions seem to be of particular importance for Bayesian learning. For example, Hilbert spaces of square integrable holomorphic functions, known as Hardy spaces as mentioned above, play an important role in the work of Trogdon–Olver [69] and Segal–Wilson [73] are useful for obtaining exact solutions of the KdV equation, as well as H_∞ control of 2-dimensional fluids. H_∞ control is an extension [84] of the Kalman filter which is of inherent interest because of its connection with game theory. Wiener [97] drew attention to eigenstates of "reproducing" kernels as being of special importance for machine learning because of their usefulness for understanding nonlinear systems with noise. The usefulness of quantum kernel spaces for machine learning has also been emphasized by Schuld *et al.* [98,99]. One class of representing kernel spaces that is of particular interest to us are the Θ-functions [53–55] whose representing kernel is the quantum propagator for a closed string [100].

One important way in which quantum mechanics differs from classical mechanics is the role played by measurements. The holomorphic states constituting Hardy spaces ("BFS states") are not orthogonal, which means they are not completely distinguishable by "von Neumann projections" [101]. However, Helstrom's theorem [102] provides a way of characterizing the information content of any quantum measurement when the measured states are not orthogonal. In a nod to quantum computing, "measurement-based quantum computing" [132] is both a mirror and an inspiration for our quantum approach to Bayesian inference. One feature that our approach to Bayesian inference shares with measurement-based quantum computing is that measurements of a momentum variable act as the control variable, guiding the system to evolve in a desirable way.

Chapter 7 caters to the notion that our quantum path integral interpretation of the innovation for the traveling salesman problem (TSP) [36] opens the door to a general quantum theory of optimal control and RL, and at the same time provides new insights into the nature of mammalian cognition. As a first step in this direction, we observe that our path integral solution of the TSP [36] can be generalized to a quantum version of the Helmholtz machine by replacing the probabilistic Ising spin degrees of freedom in the original Helmholtz machine with quantum string-like degrees of freedom [52]. In contrast with the TSP problem, where the itinerary for the salesman is regarded as fixed, the part of the quantum Helmholtz machine representing the environment is "flexible", which requires introducing methods similar to those used to solve the KdV equation, e.g. the Riemann–Hilbert method for solving the KdV or NLS equations. In our quantum version of the Helmholtz machine, the wake sleep algorithm which leads to the MDL models for input data is replaced by the stochastic version of the Feynman–Vernon [47] "influence functional". Adapting the wake–sleep algorithm to our quantum Helmholtz machine then leads to a Fokker–Planck representation for the evolution of the Bellman function for observer/controller and system/environment pieces of the Helmholtz machine. The Gaussian processes associated with this Fokker–Planck evolution correspond to quantum variations in the shape of a surface.

It is interesting in this connection how Pontryagin control, which is the deterministic limit of Bellman's theory of optimal control, emerges. Since quantum mechanics involves probabilistic predictions, one might assume that quantum mechanics and Pontryagin control would have little in common. However, many-body quantum systems have the property that under certain circumstances they can undergo a phase transition to a collective state characterized by a single space-dependent complex number referred to as the ODLRO order parameter [103]. In this state quantum fluctuations are suppressed, and the state of the system can be described as a quasi-classical Eulerian fluid with a well defined density and velocity. The precise story [104] is that a path integral description of the thermal density matrix for a gas of quantum bosons requires longer and longer paths as one approaches a quantum critical point where the speed of sound vanishes and ODLRO appears. After the appearance of ODLRO the flow of a bosonic fluid becomes very smooth and in two-dimensions describable using holomorphic functions. The control problem then resembles H_∞ control [84]. The appearance here of a path integral description of a thermal density matrix is a replay of our quantum theory of innovation [36] below the Kappin transition.

Chapter 7 provides a partial answer to the fundamental question motivating our presentation as to whether quantum theory might offer some insights into why the cognitive capabilities of the mammalian brain are in many respects better than the data analysis capabilities of state-of-the-art computers. Our penultimate result is that a quantum version of the wake–sleep algorithm for training the Helmholtz machine provides a connection between using the Riemann–Hilbert method to obtain exact solutions of integrable PDEs in terms of holomorphic functions on a Riemann surface and determining the Bellman function in feedback control and RL problems. Viewed through the lens of Kohonen's self-organizing maps [12,13], this connection may well be the long sought "explanation" for the remarkable cognitive capabilities of mammalian species. For example, comparison of a self-organizing map for the outputs of somatosensory sensors on a human hand [12,13] with multi-soliton solutions of the KdV and similar integrable PDEs reveals that there is an astonishing similarity between the two systems with respect to the role that topology plays in distinguishing different kinds of information.

Our general approach to quantum Bayesian learning makes use of the fact that the Feynman–Keldysh double path integral [47] theory of interacting quantum systems looks a lot like the Helmholtz machine. If the dynamics of interacting quantum systems can serve as a model for the recognition and model generation networks of a Helmholtz machine, our goals in this book will have essentially been achieved. In particular, the dynamics of one of the arrays is frozen in time, then the coupled arrays can serve as a model for the feedback control systems in general (cf. Fig. 2.1), or the Kalman–Bucy scheme [3]. Among the advantages of such a representation is that it allows one to use the KdV or NLS equations to determine the time-dependent reward function by choosing initial and final states for the Lax equation as the "initial conditions" for the KdV or NLS equation. The connection of the KdV/Loiuville models with Bayesian learning arises from the map between the two Hilbert spaces of holomorphic functions (which hereafter we will refer to using their eponymous name Hardy spaces).

In Chapter 8, we turn to explore the apparently deep questions relating quantum mechanics, mathematics, and the ability of mammalian cerebral cortex to create holistic 3D representations of its environment. The lifeblood of science is observation, and in an important sense cognitive science is (or ought to be) also about how the mammalian brain makes sense of observations. One key to understanding this is undoubtedly Kohonen's self-organizing maps [12,13]. Self-organizing maps have the property that they can organize sensory data in such a way that one can understand what the data mean by "visual inspection". This is perhaps a realization of the oft-quoted admonition that a picture is worth a thousand words. We are attracted to Kohonen self-organization not only because it provides a possible explanation for the way sensory neurons are organized in the mammalian cerebral cortex [107–109] but also because self-organization is relevant [108–110] to the problem of how to design arrays of artificial sensors that can efficiently fuse together different kinds of information. At the end of Chapter 8, we comment on what this means for the relationship of mathematics and theoretical physics.

A notable aspect of connecting mammalian cognition and quantum theory is that there is a very elegant quantum computational model for the seemingly effortless way mammals construct 3D representations of their environment. This model is based on the work of Julian Schwinger [111] (as a graduate student!) on the connection between quantum oscillator states and the quantum theory of angular momentum, In Chapter 8, we show how Schwinger's quantum oscillator representation for quantum angular momentum states can be used to construct a double pyramid. This construction is related to the Biedenharn–Elliott identity for the Racah 6j symbols [112]. In particular, we make use of Schwinger's oscillator representation of Wigner's triangles, together with the theory of quantum teleportation [113], to construct "by hand" a double pyramid. This construction exposes both a deep connection between the quantum angular momenta and the topology of three-manifolds and the ability of quantum wave functions to represent nontrivial knots. A perhaps deeper view of the usefulness of combining Wigner's representation for triangles together with quantum teleportation of continuous variables [113] to form holistic images of 3D objects is that this may shed light on how mammalian brain constructs holistic representation of its environment.

Many technical details necessary for a complete understanding of our presentation are relegated to Appendices. This includes a brief description of Gaussian processes and artificial neural networks, certain intractable wave scattering problems could be exactly solved by viewing the wave number as a complex number valued variable rather than just a real parameter. In a historical sense this discovery is the fundamental mathematical result underlying our presentation. In the 1950s and 1960s, the Wiener–Hopf method was adapted to the problem of describing wave propagation in a channel with a flexible boundary [116] as well as the "inverse scattering" problem in three-dimensions [117]. This application for the Wiener–Hopf method also provides the framework for Dyson's adaptive optics formalism, as well as much of our quantum approach to Bayesian inference. Appendix D is devoted to a noteworthy historical detail regarding the Riemann–Hilbert method for finding exact solutions of the KdV and NLS equations described in Trogdon–Olver [69]. Namely, the structure of the Riemann Method that underlies the T–O numerical method for solving initial value problems for the KdV and NLS equations has an interesting connection with the relationship between varieties of strongly interacting elementary particles developed in the late 1950s by Gell-Mann and Ne'eman, that was based on a novel way of generating representations of the SU(3) group known as the "Eightfold

Fig. 1.2.   Duality of optimal control and pattern recognition.

Way" [118]. The Gell-Mann–Ne'eman construction also leads to a construction [119] of solution of the KdV equation using fermionic operators.

Figure 1.2 is a cartoon illustrating how we imagine that the topics discussed in the following chapters are related to one other. Our main theme is that Bayesian inference involves two fundamental optimization principles: the Rissanen–Mumford minimum information description (MDL) principle [26] and Bellman optimization [19,20]. As noted by Todorov [19], these principles are closely related, leading to a duality between pattern recognition and optimal control/RL-type problems. In the end one can view this duality as a consequence of quantum self-organization, i.e. the tendency of quantum mechanics to favor certain types of holomorphic representations for input data and data features, particularly holomorphic representations of the Weyl–Heisenberg group related to the theory of integrable PDEs and Riemann surfaces, thus providing a plausible beginning for understanding the remarkable cognitive capabilities of the mammalian cerebral cortex.

Chapter 2

Six Fundamental Discoveries

2.1. Bayes's Probability Formula

Thomas Bayes was born in 1701 and died in 1761. He was an ordained minister in Turnbridge Wells — about 35 miles southeast of London. Although he published no scientific papers during his lifetime, his mathematical abilities must have been known to his contemporaries because he was elected a Fellow of the Royal Society in 1742. After his death, his "Essay Towards Solving a Problem in the Doctrine of Chances" was published in the *Philosophical Transactions of the Royal Society* [82]. This essay, arguably one of the most important papers in the history of science, introduced a rigorous methodology for estimating the probabilities for different possible explanations for experimental observations based on the "evidence" [5,6]. Given the revolutionary implications of Bayes's essay, especially compared with what was understood in the 18th century about the scientific method, one might have expected that his formula would have immediately been celebrated. Unfortunately, in what perhaps may be assayed as one of the most significant lapses in the post-Renaissance progress of science, it took more than 300 years after its publication for the great value of Bayes's theorem for data analysis to be fully appreciated. Fortunately, the fundamental usefulness of Bayes's formula for data analysis, Markov decision problems, and optimal control is now widely appreciated (see e.g. [6]) — even if not widely used.

The origins of Bayes's essay are not entirely understood, but it seems likely [82] that he was motivated by the earlier work of another

self-taught amateur mathematician, Thomas Simpson, on the problem of how to combine multiple observations of the positions of an
astronomical body to obtain the best estimate of the true position
of the body. Simpson introduced the fundamental notion that in
order to understand the truth behind experimental observations, one
needed to have some understanding of how the errors in observations,
i.e. the difference between the observed and true position of the body,
are distributed. In modern terminology, this would mean specifying
a probability density for the observational error. The tremendous
power of Bayes's formula lies in the fact that in a general setting it
provides a canonical approach to finding best explanation α for a set
$\{\mathbf{d}\}$ of input data, based on a set $\{\alpha\}$ of possible data models (cf. [6]).
Each explanation consists of a data model or hypothesis α together
with ancillary parameters θ for each data model, and the best explanation typically changes as data is acquired from new observations.
The canonical Bayesian prescription for data analysis is to maximize
at any particular time the *posterior probability* $p(\theta|\mathbf{d}, \alpha)$ given a par-
ticular model α and set of data $\{\mathbf{d}\}$ that has been acquired up to
that time:

$$p(\theta|d, \alpha) = \frac{p(d|\alpha, \theta)p(\theta|\alpha)}{p(d|\alpha)}, \qquad (2.1)$$

where the two factors in the numerator represent the expected likelihood for the data given a particular data model and the *a priori*
probability for that particular model. In the ideal case where the *a
priori* probabilities $P(\alpha)$ for the occurrence of various explanations α
and the conditional probability densities $P(d|\alpha)$, for the sensor data
d within each class are known, the best possible classification procedure would simply be to choose the explanation $\alpha(\theta)$ for which the
posterior probability, $P(d|\alpha, \theta)$, is the largest. Unfortunately, in the
real world one is typically faced with the situation that neither the *a
priori* probabilities $P(\theta|\alpha)$ for the various possible explanations nor
the conditional probability densities $P(d|\alpha)$ for the input data given
a particular explanation are precisely known. Therefore, one must in
general rely on ad hoc models for these probabilities in order to find
the best explanation for a particular dataset. In his comprehensive
2003 review, Mackay [6] describes the strategies behind the various
efforts to make practical use of Bayes's formula for data analysis,
and has a nice discussion as to why Bayes's prescription favors the
simplest models in the sense of Mumford–Rissanen [26].

One of the reasons for the long historical delay in making extensive use of Bayes's formula was apparently confusion as to how one could estimate the in general unknown *a priori* probability distributions. This uncertainty about the usefulness of Bayes's formula was eventually dissipated by the development of Bayesian approaches to search and control problems; where the problem of defining the prior probabilities for possible explanations is side-stepped by using the probabilistic predictions from the prior step of the search or control process as the input *a priori* probability for the next step. The problem with the initial *a priori* probability lingers, but it is often the case that the final answer is insensitive to the exact initial *a priori* probability distribution. In addition, conceptual unease with uncertainties in *a priori* probabilities has been largely erased by the notion of using a model generation or "adversarial" network to predict the probabilities for observed data.

2.2. The Wiener and Kalman–Bucy Filters

In second and third place behind Bayes's formula, we believe that data science's greatest historical advance debt belongs to Norbert Wiener and Rudolf Kalman for introducing methods for signal extraction and state estimation methods that take into account extraneous noise associated with observational errors or errors in choosing the best model to explain the observational data. Wiener's approach to signal extraction was not based directly on Bayes's formula, but instead on the difference between the correlation structure of signals and noise. In the 19th century, Adrian Legendre and Karl Friedrich Gauss initiated [82] the least squares estimation method which allows one to interpolate between examples where the explanation for the input data is known. Although there was a hint of the method of least squares in a 1722 paper by Roger Cotes, Legendre's 1805 book *Nouvelle methodes pour la determination des orbites des cometes* contained the first clear presentation of the least squares method. Legendre illustrated his method with the practical problem of determining the Paris meridian from survey data. Following Legendre's 1805 presentation, the method of least squares became widely used in astronomy and geodesy for representing data. In these early applications though it was not clear in what sense Legendre's method provided the "best" way of representing data, nor what was its relationship to probability theory. The answer to these questions was partially answered by Gauss, who showed that the method of least squares did provide a probabilistic best approximation for the data when the errors for the data could be described using Gauss's eponymous probability distribution. Gauss also introduced a recursive least squares method to carry out the calculations he undertook to locate the asteroid Ceres [82]. Unfortunately, in the presence of noise the least squares method doesn't necessarily provide the best estimate for the signal. The solution to this problem was eventually provided by Wiener [81] and Kalman–Bucy [83].

To his surprise Wiener discovered in the course of applying the least squares method to the problem of extracting a continuous time signal from noise that, at least in the case of white noise, the problem could be analytically solved by solving a certain nonlinear integral equation that he and Hopf had introduced in 1931 in connection with finding analytic solutions to certain scattering problems [61]. A simple way of understanding how this integral equation arises is provided by Yaglom's projection theorem [83]. Let us consider the Hilbert space $\mathcal{H}_x$ generated by a random process $X(t)$, where $0 < t < T$. Specifically, let us assume that $X(t)$ is a Gaussian process with a covariance matrix $R(s,t) \equiv E[X(t)X(s)]$, where XY means the dot product of vectors X and Y. The smoothing problem is: given the Hilbert space H_Y^T generated by observations $Y(t)$ of signals $Z(t)$ perturbed by white noise $N(t)$, i.e. $Y(t) = Z(t) + N(t)$, find the least squares estimator $\hat{Z} \in \mathcal{H}_x$ for $Z(t)$ that satisfies the orthogonality condition

$$E\left[\left(Z(t) - \hat{Z}(t)\right) Y(s)\right] = 0 \text{ for } 0 < s, t < T, \tag{2.2}$$

which insures that $E|Z(t) - \hat{Z})|^2$ is a minimum and $\hat{Z}(t)$ aligns with Y:

$$E[|\hat{Z} - Z|^2]] = \min_{Z \in \mathcal{H}_x} E[|Z - Y|^2]. \tag{2.3}$$

Of greatest interest is the correlation function

$$R(s,t) = \delta(t - s) + K(t,s), \tag{2.4}$$

where $K(t,s) \equiv E[Z(t)Z(s)]$ is a continuous matrix defined on $[0,t] \times [0,T]$ and a delta function representing white noise. If we now define a matrix $H(t, s)$ where $s < t$ such that

$$\hat{Z}(s) = \int_{-\infty}^{t} H(t, s)Y(s)ds, \tag{2.5}$$

then Eq. (2.2). implies that

$$K(t, \tau) = H(t, \tau) + \int_{0}^{T} H(t, s)K(s, \tau)ds. \tag{2.6}$$

A matrix H satisfying this integral equation is called the Fredholm resolvent of the covariance matrix K, and serves as a "filter" for extracting a well-defined signal $\hat{Z}(t)$ in the presence of white noise $N(t)$, as is evident from Eq. (2.6). It is the restriction to $t \leq T$ that prevents Eq. (2.6) from being easily solved using Fourier transforms. However, Wiener's observed [81] that Eq. (2.6) can be solved using Laplace transforms, with the result that the filter function regarded as an analytic function of frequency is meromorphic; i.e. it is a rational function of holomorphic functions (a holomorphic function is a smooth function of a complex number whose only singularities are zeros). This seminal discovery by Wiener is the spring from which essentially all the results described in the following flow. With respect to classical Bayesian learning, Eq. (2.2) can be regarded as the fundamental equation for stochastic estimation when covariance information for the signal (as opposed to just a model for the time dependence for the signal) is available.

Wiener's analytic solution for Eq. (2.6) did find significant applications, e.g. to antiaircraft control [19]. As it happened though, the success of the Wiener filter was eventually overtaken in importance by the Kalman–Bucy filter [83,84], which was developed independently by Richard Bucy and collaborators at the John Hopkins Applied Physics Lab and Rudolf Kalman, and jointly published by Kalman and Bucy in 1960 [83]. Like Wiener, Kalman and Bucy were interested in analyzing a time series of observations in order to make predictions about the future state of a system. As in Weiner's filter, the basic strategy for extracting signals in the presence of both measurement and system noise is to use the difference between signal and noise time correlations. In addition, by assuming that systems evolved linearly in time, Kalman and Bucy were able to reduce optimal control of the problem to numerically solving an ordinary differential equation.

Kalman and Bucy's great achievement was to describe a practical method for updating one's knowledge of the state of a system based on a linear model for the system dynamics and a control variable equal to the covariance of a Gaussian process describing the stochastic difference between the observed time history of a certain feature $z(t)$ (defined *a priori* as a linear combination of the components of the vector $x(t)$ describing the history of the system) and an underlying model value for this history. The probabilistic aspects of the observations as well as the system dynamics are taken into account by introducing white noise for theses quantities. Of course, there is a certain art to choosing the signals that are most useful in practice. (In the Indian forest described in Rudyard Kipling's famous book, the antelopes rely on the calls of monkeys and birds to discern the presence of tigers.) The great success of Kalman's model derives from the fact that given a choice for extracting useful signals from the raw data, Kalman's equations are solvable via numerical integration of an ordinary differential equation.

Kalman assumes that the input for the system consists of an exogenous perturbation $f(t)$ and a feedback control variable $u(t)$. The aim of the Kalman filter is to make use of a series of measurements $\{Y_k\}$ in order to minimize the uncertainty $\tilde{X}(t) = (X(t) - \hat{X}(t))$ in the state of the system, where $\hat{X}(t)$ is an underlying model for the system dynamics. In the discrete case, one can speak about a "gain" matrix K, which describes the gain in knowledge about the state of the system after each measurement:

$$X_k = X_{k-1} + K_k(Y_k - H_k X_{k-1}). \qquad (2.7)$$

The success of Kalman's filter also relies on the use of two Gaussian processes: (1) a noise source $N(t)$ which limits the ability of an observer to measure a signal (i.e. $Y(t) = \hat{Z}(t) + N(t)$), and (2) another GP $w(t)$ which represents an intrinsic randomness in the system dynamics. Kailath [83] introduced the designation "innovation" for $u(t)$ in recognition of the fact that it represents that part of an observation which yields information about the new state. The general scheme (Fig. 4.1) can be pictured as an interaction between an "observer–controller" and a system; e.g. a mechanical device or an "environment" (cf. the zebras and their surroundings in Fig. 2.1). The objective of the Kalman filter is that, given the GPs, $w(t)$ and

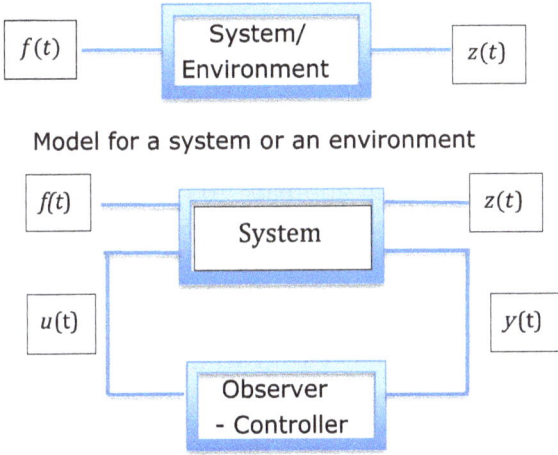

Fig. 2.1. Kalman–Bucy scheme for feedback control.

$v(t)$, minimize the mean square difference between the estimated current state of the system $X(u,t)$ and a desired target state X_T.

In cases where both the state $X(t)$ of the environment and measurements $Y(t)$ are continuous matrix functions of time, these matrices satisfy:

$$\hat{Z}(t) = H(t)\hat{X}(t), \dot{\hat{X}}((t) = F(t)\hat{X}(t) + K(t)(Y(t) - \hat{Z}(t)), \quad (2.8)$$

where the observations $Y(t)$ differ from the signal $Z(t)$ by an observational noise $\nu(t)$; i.e. $Y(t) = Z(t) + \nu(t)$, $K(t) = P(t)H' + G(t)H\hat{X}$ describes the increase in our knowledge of the system based on continuous observation of particular features of the system, and it is assumed that all the coefficient matrices are all known functions of time. Here, H_k is a matrix which defines the "features" $\{Z_k\}$ of the environment which are of greatest interest to designated controllers. It is a result of combining observation of a system (or 'environment' in the case of RL) with the linear dynamical model, Eq. (2.3), for the system that the controller hopes to gain enough information about the state or environment to take effective corrective actions.

The time-dependent covariance $P(t) = E\tilde{x}(t)\tilde{x}(t')$ for the error $\tilde{x}(t) = (x(t) - \hat{x}(t))$ can then be found [118] by numerically solving a ordinary nonlinear differential equation:

$$\dot{P} = -PH^T R^{-1} HP + FP + PF + Q, \quad (2.9)$$

where v and w represent observation and system noise with covariances Q and R.

The Kalman–Bucy "filter" garnered much praise in the 1960s as a result of its successes for spacecraft control [3]. However, the assumption that state variables evolve linearly in time is rather restrictive, and the Kalman approach to control theory was eventually supplanted as a general approach to optimal control by Richard Bellman's "dynamic programming" formalism [20].

2.3. Bellman's Dynamic Programming Approach to Optimal Control

The filters of Wiener and Kalman–Bucy might reasonably be regarded as the seminal springs for machine learning. However, we are indebted to Richard Bellman [20] for having been the first person to introduce a quantity that one wants to optimize with feedback control (although in a certain sense this was implicit in the 14th century work of William of Ockham [85]). Bellman's great contribution was to introduce a performance index $V(t)$ for the efficacy of feedback control together with a recurrence relation for determining its value. Bellman referred to this performance index as a "cost-to-go". As in classical mechanics where the object is to minimize the Maupertuis action [88], the object of optimal control is to minimize the Bellmann cost function. (One thing that is confusing about the optimal control literature is that $V(t)$ is often referred to as a "value" function for a control or RL strategy, even though the object is to minimize its magnitude.) Bellman's dynamic programming approach to optimal control [20] is based on the idea that the optimal cost-to-go should not depend on whether one carries out the optimization all in one campaign or in two steps, leading to a recursion relation for a "cost function" $V(t)$:

$$V(T) = \frac{\min}{[t, t_1]} \left\{ \frac{\min}{[t_1, T]} \left[\int_t^{t_1} l(X(\tau), U(\tau), \tau) d\tau \right. \right.$$

$$\left. \left. + \int_{t_1}^T l(X(\tau), U(\tau), \tau) d\tau \right] \right\}. \qquad (2.10)$$

The main burden of Bellman's approach to feedback control is finding practical ways of solving Eq. (2.10). To a significant extent the frontier of data science is defined by finding better ways of calculating the loss function $l(X, U, t)$ for Bellman's cost function, especially in complex situations. Perhaps the most interesting aspect of Eq. (2.10) is the important role played by the "loss" function in its connection with information theory.

Although Bellman didn't initially phrase it this way, it was eventually realized that the loss function represents the rate at which information about the optimal path is accumulated. Worth mentioning in this respect is a 2011 paper [28] prompted by the Monte Hall controversy [5], which clarified that the rate of accumulation of Shannon information during a Bayesian search can be assayed by the change in Bellman's cost function (where the change in information is identified with the negative of the change in Shannon entropy). This information theory interpretation for Bellman's value function is consistent with Todorov's interpretation [19] of the negative exponential of the Bellman value function as the "backward filtering" probability $p(y(n), \ldots, y(N)|x(n))$ for obtaining a series of measurements $\{y(i), i = n, n + 1, \ldots, N\}$ in the future, given the current state $x(n)$ of the system. Thus, any attempt to provide a quantum interpretation for Bellman's approach to optimal control would necessarily require including a model for how future measurements will affect one's understanding of the state of the system. This echoes the "time-symmetric" formulation of quantum mechanics [18], and contrasts with classical mechanics where one is only allowed to specify an initial, or alternatively, a final state for a system.

Curiously, Bellman's invention of dynamic programming followed not long after Feynman's development of his path integral approach to quantum mechanics for his PhD thesis at Princeton [32]). The fact that Bellman was a PhD student of Soloman Lefshetz at Princeton in the early 1950s certainly allows for the possibility that Bellman was aware of Feynman's ideas at the time he developed his dynamic programming algorithm (although Feynman left Princeton during WWII and went to Cornell University after the war). One of our main threads in the following will be exploring the relationship between Bellman's dynamic programming algorithm and Feynman's path integral approach to quantum mechanics.

2.4. Feynman's Path Integral Approach to Quantum Mechanics

In the seminal papers of Dirac and Heisenberg, Born, and Jordan [42], it was assumed that the classical equations of motion could be written in the form given by Hamilton's equations of motion, where the time derivatives of the position q and momentum p variables are written in the form

$$\dot{p} = -\frac{\partial H}{\partial q}, \quad \dot{q} = \frac{\partial H}{\partial p}, \tag{2.11}$$

where $H(q, p)$ is a function, known as the Hamiltonian, describing the physical system. In the matrix mechanics of Heisenberg [40], the classical quantities q, p, and H are represented as matrices. In addition, the variables q and p describing the state of the system satisfy the commutation relation $[q, p] = i\hbar$, where $\hbar$ is $1/2\pi$ times the constant that Planck introduced in connection with his quantum treatment of thermal radiation. Indeed, the first estimates of the value of $\hbar$ were obtained by comparing Planck's theory of thermal radiation [40] with experimental measurements of the spectrum of infrared radiation from ovens. Although the canonical Eqs. (2.11) are very simple, a perhaps more elegant way to derive the equations of motion of classical mechanics is to make use of the Principle of Least Action. In this way, classical mechanics is derived from an optimization principle. The quantity that is minimized, the "action", is defined as the time integral of a Lagrangian function:

$$S(T) = \int_0^T L(p, q)dt, \tag{2.12}$$

where

$$L = p\dot{q} - H(q, p). \tag{2.13}$$

Although using a Lagrangian function of q and dq/dt rather than a Hamiltonian function of q and p might seem to be a trivial difference, it turned out that using the Principle of Least Action as the starting point for formulating quantum mechanics made a profound difference.

Following Dirac's lead [35], Richard Feynman investigated in his PhD thesis [32] what role the classical principle of least action might play in quantum mechanics. As a result, he was led to an entirely new way of formulating quantum mechanics. Rather than beginning with the Schrodinger equation which had previously been the starting point for quantum mechanical calculations, Feynman focused on the "propagator" $K(x_b, t_b; x_a, t_a)$ which transforms the Schrodinger wave function in the position representation as a function of time:

$$\psi(x_b, t_b) = \int K(x_b, t_b; x_a, t_a)\psi(x_a, t_a)d^3x_a. \qquad (2.14)$$

Feynman began by considering the form of (2.14) in the limit where t_b is very close to t_a. Following Dirac, Feynman assumed that when t_b is infinitesimally close to t_a, what is known in classical mechanics as a contact transformation, Eq. (2.14) becomes

$$\psi(x_b, t_a + \varepsilon) = \frac{1}{A(\epsilon)} \int e^{i\epsilon L(\frac{x_b - x_b}{\epsilon}, x_b)}\psi(x_a, t_a)d^3x_a, \qquad (2.15)$$

where $L(\dot{x}, x)$ is the classical Lagrangian function (2.12) and $A(\epsilon)$ is a normalization factor, which is required in the transition from classical to quantum mechanics and whose exact form depends on the specific form of the Lagrangian. By concatenating infinitesimal transformations of the form (2.15), Feynman arrived at his path integral for the propagator; i.e. the quantum amplitude for going from x_a to x_b:

$$K(x_b, t_b; x_a,) = \int_a^b e^{i\hbar S[b.a]}D[x(t)], \qquad (2.16)$$

where $S[b.a]$ is the classical action (cf. [47]) going from x_a to x_b, and $D[b, a]$ denotes a sum over all paths leading from x_a to x_b. For a free particle with mass m in one-dimension, this propagator takes the simple form [47]:

$$K(x_a, t_b; x_a, t_a) = \frac{1}{A(t_b - t_a)} \exp \frac{im}{2\hbar} \left[\frac{(x_b - x_a)^2}{t_b - t_a} \right], \qquad (2.17)$$

where $A(\tau) = [m/2\pi i\hbar(t_b - t_a)]^{1/2}$. The exponent of the exponential factor is just $(i/\hbar)$ times the classical action for a free particle. If instead of assuming that the particle started at a definite position, we assumed that the particle started with a definite momentum, then the amplitude for the particle to arrive at a position x is

$$\psi(x,t) = \frac{1}{A(t)} e^{\left[ipx - i\left(\frac{p^2}{m}\right)t\right]/\hbar}. \tag{2.18}$$

The exponent of the exponential will be minimized when $x = (p/m)t$, which is just the classical equation of motion for a free particle. The probability of finding the particle at a point x is the absolute square of the r.h.s of Eq. (2.9) is $1/|A|^2$, which depends on time but is independent of x. This reflects the fact that in quantum mechanics when the momentum is known with certainty, the position variable is completely unknown. This complementarity would seem to be unhelpful from the point of view of classical physics, but from the perspective of data analysis it will turn out that this complementarity is very useful.

2.5.   Quantum Solution of the Traveling Salesman Problem (TSP)

The primary inspiration for our presentation is the observation [36] that quantum mechanics provides a simple solution to the problem of finding the order in which a salesman visits cities, and provides a canonical example of the type of topological obstruction that often attends Bayesian model selection [36,78]. As was emphasized in Kailath's review of linear noise filters [83], a crucial ingredient needed for these filters is the Fourier transform of the anti-causal correlation function for the signal plus noise regarded as an analytic function of frequency (cf. Appendix B). The crucial discovery that we would like to highlight is the discovery [36] that the appearance of holomorphic functions (rational functions of smooth functions of a complex number whose only singular points are zeros) in pattern recognition and feedback control has a simple quantum mechanical interpretation in terms of Feynman path integrals; at least in the case of the TSP. The solution to the TSP we describe below also illustrates why meromorphic functions (ratios of holomorphic functions) naturally arise in data analysis problems involving combinatorial optimization. Of course, it is not always true that data analysis involves combinatorial optimization, but what sets the mammalian brain apart from the brains of other animal species is that the mammalian brain has some capability of dealing with ambiguities in the interpretation of sensory data, which necessarily involves [78] combinatorial optimization.

Our introduction of the term quantum self-organization in connection with the TSP is a pointer to the relationship between our solution for the TSP and the appearance of holomorphic functions in Kohonen self-organization of sensory data [12]. Our discovery was prompted by the Durbin–Willshaw elastic net method [1] for finding solutions to the TSP. As the name suggests, this involves adding to the locations of the cities to be visited, indicated by the round dots in Fig. 2.1, a trial "itinerary" for the salesman, the square points in Fig. 2.2, and then connecting all the points to each other by springs. The lengths $d(i, \mu)$ of the springs connecting the nodes where the salesman is assumed to stop to the actual location of cities that are to be visited is the "innovation" for the Durbin–Willshaw method; i.e. the distance between the actual locations of the cities and a model for the salesman's itinerary.

In the Durbin–Willshaw approach [1], the TSP is solved by the simple expedient of connecting the initially randomly placed movable nodes with elastic strings, and the cities to be visited by nonlinear strings, and allowing the system to relax to the lowest energy state using gradient descent dynamics for an energy functional [1]:

$$E[\{w_i\}] = -\sum_{\mu} \log \left\{ \sum_i \left[\exp\left(\frac{|\xi^\mu - w_i|^2}{2} \right) + \frac{K}{2} |w_{i+1} - w_i|^2 \right] \right\}.$$

(2.19)

When the locations $\{\xi^\mu\}$ of the square points in Fig. 2.2 are not too far from round points w_i, this approach gives a satisfactory solution for the TSP via gradient of the energy functional (2.19). What makes the traveling salesman problem especially interesting from the point of view of using quantum mechanics to solve optimal control and RL problems is that the term in Eq. (2.19) involving the difference in the positions of the round and square points can be replaced [36] by a Feynman path over all paths marked with squares.

$$K(t - t_0) = \int \exp\left(\frac{i}{\hbar} \int_{t_0}^{t} \left(\frac{m}{2} |\dot{x} - \mathbf{v}(t)|^2 \right) dt \right) Dy(t'), \qquad (2.20)$$

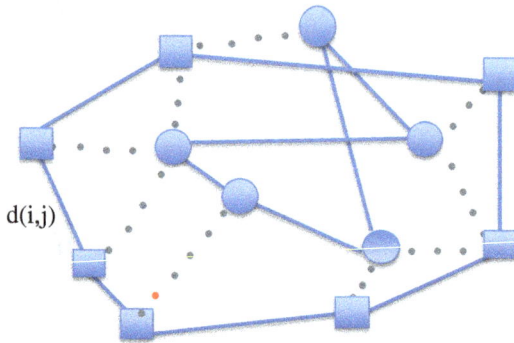

Fig. 2.2.   Durbin–Willshaw setup for solving the traveling salesman problem.

where the classical velocity $v(x)$ is defined for all x by the actual motion of the salesman, and $y(t) = x(t) - x_{cl}(t)$ is the deviation of the Feynman path from the salesman's itinerary.

Equation (2.19) together with Fig. 2.2 implicitly illustrates one aspect of the Bayesian model selection problem that is particularly troublesome; namely, connecting data points with models will in general involve a topologically non-trivial planar graph (i.e. a planar graph where at least two lines defining the graph cross one another). If one changes the order in which the cities are visited, then in general both the solid line marking the salesman's path and the dashed line will cross one another. This makes improving the model for the salesman's using Markov chain Monte Carlo regression methods essentially intractable (see e.g. [79]). However, because of the general mathematical equivalence of topologically nontrivial planar graphs and topologically simple paths on Riemann surfaces [80], the regressions that are ill-defined for planar graphs can be carried out on a topologically nontrivial surface.

In an almost obvious way, the Durbin–Willshaw setup can also be interpreted as a control problem by interpreting the x and y coordinates of the square dots as estimations of a time-dependent vector $X(t)$ describing the evolution of the state of a system in phase space (viz. the evolution of position and velocity variables $(x, \dot{x})$ for a self-driving car). In this interpretation, the round points represent an underlying model $\{\hat{z}(t)\}$, for how these variables vary with time. The distances $d(i, \mu)$ in Fig. 2.2 corresponds to Kailath's "innovation

Gaussian process" [83], which describes the difference between observations and the predicted state of the system used in the Kalman–Bucy filter method [3,83] for optimal control. The terms in Eq. (2.18) involving just the estimated positions $\{w_i\}$ measure the area swept out by the elastic string; so we see here the emergence of a connection between optimization of Bellman's value function for system control [20] and the Nambu action principle in string theory [100].

Elevating the true and estimated locations of the cities to be visited in the Durbin–Willshaw setup onto a Riemann surface allows the cities to be connected by a simple non-self-intersecting path with only modest changes in the direction of motion after visiting each city. A crucial observation is that this in turn allows one to obtain Kailath's innovation, defined as the difference between estimated and model histories for the salesman's path, as a smooth function on the surface without the topological obstructions to finding the optimal itinerary for the salesman that arise if it would be needed to examine all permutations of the order in which the salesman visits the cities. If a topologically nontrivial itinerary for the salesman defined on a plane is lifted to Riemann surface with sufficiently large genus, i.e. number of "donut holes" (this is always possible [80]), then the use of Monte Carlo relaxation to find a good approximation to the optimal path, even with a large number of cities, would become straightforward.

Chapter 3

Ockham's Razor

3.1. Bayesian Searches

The modern-day impetus for the development of information-based Bayesian learning was provided by the WWII problem of locating submarines [122]. A general Bayesian search is defined by a sequence of decisions $U_N \equiv \{x_{i+1} - x_i, i = 1, \ldots, N - 1\}$ leading to a sequence $X_N \equiv \{x_i, i = 1, \ldots, N\}$ of compact subsets in a d-dimensional space that will be interrogated for the presence of the object at locations x_i. The possible locations x^* of this object are indexed by a variable $\mu = 1, \ldots, M$. The interrogations result in a sequence of measurements $Y_N \equiv \{y_k, k = 1, \ldots, N\}$ that are intended to determine at each step n whether the sought-after object is present in a localized setting. In reality the result of the observation y_n at step n only determines probabilistically whether x_n is the location, i.e. $x_n = x^*$. Based on the collection Y_n of measurements gathered during the first n steps of a search and the probability density $p_{n-1}(x)$, one can estimate the posterior probability density $p_n(x) \equiv p_n(x = x^* | X_n, Y_n, p_{n-1})$ that at step n the sought after object is located at location $x = x^*$ is

$$
\begin{aligned}
&p_n(x^* = x | Y_n, U_{n-1}, p_{n-1}) \\
&= \frac{p(Y_n | U_n, X^* = x, p_{n-1}) p_{n-1}(x^* = x | Y_{n-1}, U_{n-1})}{p(Y_n | U_n)}.
\end{aligned} \quad (3.1)
$$

Of course, future decisions are needed only if the object isn't detected.

A simple real-world search problem that can serve as an introduction to general problem of Bayesian model selection is the problem of searching in two-dimensions for the location of an object X^* located at a location, say x^*. For example, one could be searching for the location of an airplane that crashed in some region of the Atlantic Ocean. This classic Bayesian search is characterized by a series of compact areas $\{A_n\}$, and associated answers to the question "Is X^* located inside A_n?" In order to inject an element of realism, it is further assumed that the answers to the questions regarding the location of the object are not communicated in an unambiguous fashion, but only as measurement results $\{Y_n\}$ which are probabilistically related to the true answers $t_n = 1(x^* \in A_n)$ or $t_n = 0(x^* \notin A_n)$. We are interested in the posterior probability densities $p_n(x^*|D_{n-1})$ for finding the object in the vicinity of location x^* after n-steps, given a dataset $D_{n-1} \equiv \{(A_1, Y_1), \ldots, (A_{n-1}, Y_{n-1})\}$ that has been gathered regarding the location of the object prior to step n. The theory of Bayesian searches is based on the use of Bayes's formula, Eq. (1.1), to evaluate at each step of the search the conditional probability density $p_n(x^*|X_n, Y_n)$. In the context of our search problem, the explanation α being sought is the location x^* of X^*, while the input data $\{d\}$ consists of the sequence $\{(A_1, Y_1), \ldots, (A_{n-1}, Y_{n-1})\}$ of prior choices for search locations and measurement results. The prior probability $p(\alpha)$ for finding X^* in A_n is $p_{n-1}(A_n) \equiv \int_{An} p_{n-1}(x|D_{n-1}) dx$; i.e. the probability that X^* is located in A_n given the information gathered up to the time of the previous step. The conditional probability $P(d|\alpha)$ is the probability for obtaining a response y if the search volume is A_n and X^* is located at x^*:

$$P(Y_n = y|A = A_n, x^*) = \begin{cases} f_1(y), t_n = 1 \\ f_0(y), t_n = 0. \end{cases} \tag{3.2}$$

The probability distributions $f_0(y)$ and $f_1(y)$ for the Y_n values will in general be different in the two cases $t_n = 1(x^* \in A_n)$ or $t_n = 0(x^* \notin A_n)$, and depend on n, but as a simplification, we assume that these probability distributions are independent of n. With these assumptions Bayes's formula for the posterior probability density distribution for the location of X^* after n search steps $P_n(x^*|D_n) \equiv p_n(x^*)$ becomes

$$p_n(x^*) = \frac{P(Y_n = y|A = A_n, x^*)p_{n-1}(x^*)}{f_1(y)p_{n-1}(A_n) + f_0(y)(1 - p_{n-1}(A_n))}. \tag{3.3}$$

This search problem corresponding to (3.3) can be visualized as a tree of observations and decisions.

Since the results of our efforts are expressed as probabilities, it is natural to inquire how much information has been gained about the location of the object after N steps of choosing subsets of R^d and observing whether the object lies within the chosen volume. A natural measure of the information gathered after n-steps is minusing the Shannon entropy:

$$H(p_n) = -\int p_n(x) \log p_n(x) dx, \qquad (3.4)$$

where the integral extends over all space and the log is base 2. The Shannon entropy, Eq. (3.4), is a measure of the progress for Bayesian searches, optimal control, and reinforcement learning. All three of these types of machine learning problems can be characterized as the problem of finding a policy for choosing the sequence of actions so that the Shannon entropy $H(p_N)$ is minimized after N-steps. Bellman's optimization of this cost function introduced in his paper on dynamic programming [20] minimizes at each step the entropy (3.4):

$$V(p, n) = -\min_\pi E_Y[H(p_N|p = p_n], \qquad (3.5)$$

where the index π denotes a particular choice for $\{A_1, \ldots, A_{n-1}\}$. It turns out that one can also rephrase search optimization as the problem of finding the choice for A_n, given a previous choice for $\{A_1, \ldots, A_{n-1}\}$ that leads to the largest expected decrease in the Shannon entropy. As shown in an important paper by Jedynak, Fraser, and Sznitman [28], the optimal solution for Bayesian search problems can also be obtained by optimizing the differential increase $\Delta I(p_n)$ in the information regarding the location x^* of the object X^* when going from step n to $n+1$. As shown in [19], the expected increase in information is

$$\Delta I(x^*, Y_n) = H(p_n) - E_{Y_n}[H(p_{n+1})|A_n = A, p_n], \qquad (3.6)$$

where the second term on the r.h.s is the expected information available after step $n+1$ given the choice $A_n = A$ assuming that prior information is information contained in the the posterior probability density $p_n(x)$. The l.h.s of (3.6) also represents the mutual information between the conditional distributions for x^* and Y_n. This mutual information can also be written in terms of the information contained in the probability density for Y_n:

$$\Delta I(x^*, Y_n) = H(Y_n|A_n = A, p_n) - E_{Y_n}[H(Y_n|x^* A_n = Ap_n)]. \quad (3.7)$$

Using Eq. (3.5), this can also be written in the form

$$\Delta I(x^*, Y_n) = H[f_1 p_n(A) + f_0(1 - p_n(A)]$$
$$- H(f_1)p_n(A) + H(f_0), \quad (3.8)$$

where $p_n(x)$ is the posterior probability density for finding x^* near location x, and $f_0(y)$ and $f_1(y)$ are the probability distributions for the y_n values in the two cases (0) x^* is not near to x or (1) x^* is near to x. The right side of (3.8) is a concave function of $p_n(A)$, and therefore has a unique maximum as a function of $p_n(A)$ at say $p_n(A) = p^*$. A central result of Bayesian search theory is that the optimal search strategy is to choose the sequence $\{A_1, \ldots, A_n\}$ in such a way that each step $p_n(A) = p^*$. With this choice for $p_n(A)$, the information regarding the location of X^* increases by a constant $C^* \equiv \max \Delta I$ at

$$E[H(p_{n+1}|)A_n] = H(p_n) - C^*. \quad (3.9)$$

After n-steps out of total of N-steps the value of the optimal search can be expressed as

$$E[H(p_{n+1})|A_n = A, p_n] = H(p_N) - (N - n)C^*. \quad (3.10)$$

Thus, we see that the optimal search strategy is characterized by constant information gain at each stage of the search. By analogy with communication theory, the constant C^* is sometimes referred to as the channel capacity for the search. What is especially noteworthy though is that (3.10) is essentially the same as the result obtained from Bellman's recursion relation for the cost function in his dynamic programming approach to optimal control.

If one defines an optimal search strategy π_n as the choice of locations that at each step minimizes the Bellman cost function:

$$V(p, n) \equiv \min_\pi E_{\pi_n}[H(p_n)|p_n = p], n = 0, \ldots N, \quad (3.11)$$

where π_n is choice for a sequence $\{A_1, \ldots, A_n\}$, then it follows from the theory of Markov processes that $V(p, n)$ satisfies the recursion relation [20]

$$V(p, n) = \min_A E[V(p_{n+1}, n + 1)|A_n = A, p_n = p], n < N. \quad (3.12)$$

This equation is essentially identical with Bellman's original dynamic programming relation [20]. One way of showing that the strategy $p_n(A) = p_n^*$ is optimal is to use the recursion relation (3.11) to show that this choice allows one to attain the minimum in (3.12). It follows from Eq. (3.12) that the universal strategy for Bayesian searches is that one attempts to optimize the search by maximizing at each stage of the search the decrease in the Shannon entropy $H(p_n) = -\int p_n(x) \log p_n(x) dx$ associated with the posterior conditional probability $\pi_n(x)$ for finding the object at various locations. Thus, the optimal search strategy can be characterized as the problem of finding a policy for choosing a control sequence so that the amount of Shannon information gathered at each step is maximized.

The explicit dependence of the state of a system or environment on the history of control actions can be exhibited by recursive use of formula 3.12, yielding

$$p(x_N = x^*|U_{N-1}, Y_N)$$
$$= \frac{\int Dx_k \prod_{k=1}^{k=N} p(Y_k|U_{k-1}, x^*, p_{k-1}) p_1(x^* = x)}{p(Y_n|U_n)}. \quad (3.13)$$

In a completely analogous manner to the way the posterior probability for a Bayesian search $p(x|D)$ was obtained by integration over all possible values of an interpolation function for input data labels, the posterior probability that after $N + 1$ steps the new state will be $x_N + \Delta x_N$ can be obtained by integrating over an interpolation function $U(x)$ for the controls

$$p(\Delta x|D_{N-1}, x_N) = \int DU(x) p(\Delta x|U(x), x_N) p(U(x)|D_{N-1}). \quad (3.14)$$

Equation (3.14) by itself doesn't determine an optimal control or RL strategy. In order to achieve this, one must introduce an optimization principle like Eq. (3.12). This brings us back to Bellman's value function Eq. (3.11), where π_n is "policy" for choosing a sequence of control decisions $U_M \{u_n\}$. It follows from the theory of Markov processes that $V(p, n)$ satisfies the recursion relation Eq. (3.12), which is equivalent to Bellman's original dynamic programming formalism. Unfortunately, as with Bellman's original dynamic programming equations, this equation is difficult to solve.

3.2.   A Tale of Two Costs

Although the goal of Bayesian pattern recognition is the optimization of the posterior probability defined in Bayes's formula Eq. (1.1), a widely used approximation is choosing the pattern that finds the best explanation for the data based on the product of $p(\alpha)$ and $p(d|\alpha)$. This approximation, known as the ML method [6], focuses on maximizing the following "likelihood function":

$$L(\alpha) = -\log[p(d|\alpha)\,(p(\alpha)].\tag{3.15}$$

The maximum likelihood method is often quite useful for what MacKay calls "the first level of inference" where one assumes that one has a model for the data that represents the underlying truth, and the data analysis task is to find the parameters for the model that provide the best fit for the given set of input data. However, in situations where different types of explanations for a particular dataset are possible, e.g. the dilemma faced by the animals in the wild trying to sense whether a predator is near, a more sophisticated method must be employed. Finding a path to a more sophisticated approach to problems such as those faced by animals in the wild will be one of our main focuses for the remainder of the book. (A better method is apparently available to mammals thanks to evolution, but alas it is not yet available to data science.) At the level of Bayesian model selection, one is often faced with the task of choosing from possibly a multitude of models $\{\mathcal{H}_i\}$ which are all possible *a priori* explanations for the data, the one that is the best fit for observations. The posterior probability for the correctness of a model is proportional to the product of the evidence of the model and the prior probability of the model:

$$P(\mathcal{H}_i|D) \sim P(D|\mathcal{H}_i)P(\mathcal{H}_i).$$

If all the *a priori* probabilities are roughly the same, then MacKay's prescription is to turn to the evidence $P(D|\mathcal{H}_i)$ in order to rank the plausibility of each model. As in the ML method, the relative likelihood of two different models is

$$\frac{P(\mathcal{H}_1|D)}{P(\mathcal{H}_2|D)} = \frac{P(D|\mathcal{H}_1)}{P(D|\mathcal{H}_2)}\frac{P(\mathcal{H}_1)}{P(\mathcal{H}_2)}. \tag{3.16}$$

The ratio of prior probabilities on the r.h.s of Eq. (3.16) allows one to input one's personal judgement regarding the relative elegance or simplicity of the two models. However, MacKay [6] points out that the ratio of the evidence factors in Eq. (3.16) allows one to assay the relative simplicity of two models in a way that is independent of subjective judgments. Moreover, the evidence factor for any model can be estimated from the way the model parameters needed for the given dataset are distributed relative to the ML value for these parameters:

$$P(D|\mathcal{H}_i) \cong P(D|w_{ML}, \mathcal{H}_i)P(w_{ML}|\mathcal{H}_i)\sigma_{w|D}, \tag{3.17}$$

where $\sigma_{w|D}$ is the width of the distribution of the model parameters needed for a given dataset. The product of the last two factors on the r.h.s of Eq. (3.23) were christened "Occam's Razor factor" by MacKay [6]. (In the literature on medieval science (see e.g. [85]) the place where William, the discoverer of this principle, lived is spelled Ockham, which is the spelling we have adopted.)

In contrast with the ML method [6], the necessity for considering many possible models or even a single model with multiple parameters' as an explanation for a pattern recognition or state estimation problem can render model selection intractable — at least in real time — with conventional computational resources. A simple way to visualize the extra complexity introduced by multiple models is provided by the linked orbit problem: instead of just determining the orbital parameters of individual astronomical objects, suppose that there is ambiguity in the observations as to how the objects in a set of objects should be associated with optical observations of orbiting objects. The likelihood function for this problem has the form (3.21). For example, suppose we want to determine the orbital parameters of N objects distributed among M orbits. Unfortunately, in this case the parameters linking objects to orbits are not readily determined as part of the observational input. Assuming that a set of observations are statistically independent, the likelihood function for a model where the N objects are assigned a particular set of orbits specified by linking parameters k_m^n that take values in $\{0,1\}$ has the form [78]

$$p(D|k_m^n) = \sum_{m=1}^{M} p(D|\theta_m, k_m^n)p(\theta_m|k_m^n), \qquad (3.18)$$

where data consisting of a series of distinct observations of the N objects and the θ_m are parameters for the orbit indexed by m, and point out that introducing parameters like k_α^i reduces a model comparison problem to a parameter estimation problem in a "product space" of model parameters θ and model selector parameters k. In principle, improved strategies for Markov Chain Monte Carlo regression [79] might help, although in practice this would be impractical if N and M are large, which serves as a reminder of the difficulties one can encounter when the number of models becomes exponentially large.

One way to view such problems is to imagine that each model α in Eq. (3.15) is associated with a set of parameters Θ. The free energy defined in Eq. (1.2) can then be written as

$$F = \sum_{\alpha} L(\alpha) = -\sum_{\alpha,\theta} \log[p(d|\alpha)\,(p(\alpha|\theta)], \qquad (3.19)$$

where $L(\alpha)$ is the likelihood function defined in Eq. (3.21). The exact answer for the best model is still given by maximizing the probability defined in Eq. (3.23), and this is equivalent to minimizing the free energy of the avatar physical system:

$$F(x) = \sum_{\alpha} \{E_\alpha P(\alpha) - (-P(\alpha)\log P(\alpha))\}. \qquad (3.20)$$

If instead of the true probability distributions $P(\alpha)$ and $P(x|a)$ one uses model probabilities $P(\alpha;\theta)$ and $P(x|\alpha;\theta)$ depending on parameters θ to calculate an approximate probability distribution $P(\alpha;\theta)$ for different classifications of a dataset x, then Eq. (3.3) will no longer necessarily be satisfied and the free energy $F(x,\theta)$ calculated using the distribution $P(\alpha;\theta)$ will in general differ from the true free energy. The advantages of minimizing the free energy (3.19) vs the ML method are captured by MacKay's Occam Razor factor.

Unfortunately, minimizing the free energy — or equivalently maximizing the evidence factor in Eq. (1.1) — using conventional Monte Carlo regression techniques is in general frustrated if there are *a priori* numerous plausible models for a set of input data. As examples are problems where M objects are distributed among N locations. In contrast with the simple Monty Hall problem considered in the last section, these types of problems are often intractable with conventional computational resources. Assuming that a set of observations are statistically independent, the likelihood function for a set of data can be written [120]:

$$P\left(\{d_n\}_{n=1}^{n=N}|\{\theta_\alpha\}_{\alpha=1}^{\alpha=M}\right) = \prod_{n=1}^{N} \prod_{\alpha=1}^{M} [P(d_n|\Theta_\alpha)]^{k_m^n}, \qquad (3.21)$$

where k_m^n in Eq. (3.21) are not readily determined as part of the observational input.

Determining MacKay's Occam's razor is also closely related to the problem of constructing "adversarial" networks, i.e. devising an algorithm which will generate authentic looking input data given a suitable choice of network parameters. The construction of adversarial networks that are of practical use in general settings is currently an active area of research in the data science community. Here will focus on the approach of Neal, Hinton, *et al.*, known as the Helmholtz machine [8,9].

3.3. Hidden Factors and the Helmholtz Machine

We now turn to the problem of selecting data models when hidden factors are important. Hidden factor analysis [20] is a type of Bayesian inference that is of considerable intrinsic as well as practical interest. For example, factor analysis is widely used for optimizing manufacturing processes and designing experimental trials of drugs [7,29]. One common reason for the proliferation of models in practice is the existence of hidden variables, i.e. variables that affect the current and future state of a system or environment, but are not explicitly recognized as input data or an external influence. In economic terms, the hidden variable problem can be very important, e.g. hidden factor analysis is widely used to design experiments to test the efficacy of experimental drugs.

Linear hidden factor analysis problem is usually formulated by regarding the hidden factors as a linear matrix equation of the form [122]

$$x = gy + \nu, \tag{3.22}$$

where x is a vector of N measurable quantities and y is a vector of hidden "factors". These are the unrecognized factors that for example could affect the reliability of a product or the effectiveness of a drug. The k x m matrix g contains the parameters for generating the hidden factors from input data. The learning problem is how to estimate the effect of the hidden factors on the results of experiments. Indeed, if in addition to the familiar state variables X_N hidden factors are involved, it is mathematically inconsistent to just include the visible variables in a Bayesian analysis of the likelihood of an observed outcome. Instead, one must maximize the information value of the evidence including possibly many hidden factors:

$$C(\{Y_N\}) = -\sum_{n=1}^{n=N} \log[p(Y_n|U_n)], \tag{3.23}$$

Initial attempts to solve hidden factor problems were based on the ad hoc assumption that the outcome is a linear or quadratic function of the hidden parameters, and this type of factor analysis is still widely used, for example, in drug testing and industrial process optimization. The importance of factor analysis invited the development of more rigorous methods for solving these types of problems. In 1977, an algorithm called the expectation maximization (EM) algorithm [29] emerged that provided a systematic approach to all types of model selection problems, including problems with hidden factors. The hidden variables are typically assumed to be Gaussian random variables, while the EM algorithm proceeds in two steps: (1) An E-step that finds the probability distributions for the hidden variables based on current estimates for the model parameters, and (2) an M-step that updates the g vector in Eq. (3.23) as well as the variances of the y components using the loglikelihood for the observed data. Unfortunately, the matrix operations involved in using the EM method don't appear to be a plausible model for learning in the brain. This led Dayan, Hinton, Neal, and Zemel to introduce a statistical mechanics-like model [8] for how the human brain might construct models for sensory data, which they called the "Helmholtz machine".

In contrast with the much simpler problem of choosing the parameters for a single GP, the 1994 Helmholtz machine papers of Hinton, Neal *et al.* [8,9] were the first papers to provide a definite scheme as to how one might construct an minimum description length (MDL) data model that takes into account hidden factors. (The name Helmholtz machine is a tip of the hat to Helmholtz's intuition that the human brain can estimate probabilities.) In the expectation maximization method by a gradient descent method. They were able to completely side-step the usual problem of *a priori* defining a framework for data models by using the same structure for the "generative" network as the "recognition" network, and in alternating epochs using the excitations in one network to modify the parameters in the other network. This procedure, known as the "wake–sleep" algorithm [9], had the miraculous properties that it could simultaneously be interpreted as the minimization of the free energy of the two networks regarded as physical stochastic systems, and deliver a probabilistic representation for the data that satisfied the Munford–Rissanen MDL principle. Furthermore, in the MDL limit the conditional probabilities connecting observed data and model parameters automatically satisfied Bayes's formula Eq. (1.1). This symmetry between the recognition and model generation networks provides the inspiration for much of what follows.

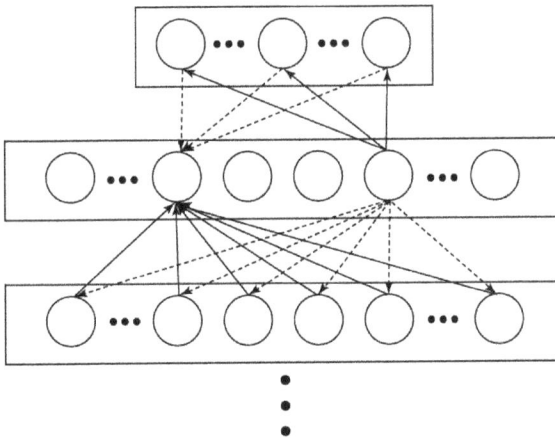

Fig. 3.1.   Helmholtz Machine recognition + data generation networks [8].

Use of the wake–sleep algorithm to solve practical problems has been severely limited because of the difficulties of calculating conditional probabilities with existing conventional computers. One of our goals in this book is to inquire whether there is a quantum resolution of the Helmholtz machine difficulties. As will be discussed in Chapter 7, there are reasons for thinking that the quantum dynamics of two interacting oscillator arrays to representing the X_N and U_N variables might provide an alternative framework for implementing a Helmholtz machine-like approach to hidden factors and model selection.

Despite their practical successes, deterministic back-propagation neural networks have the drawback that in general they neither calculate the probability that the explanation they find for the input data is correct, nor provide a guarantee that the solution is stable against small changes in the input data. Moreover, the learning algorithm for layered neural networks (see Appendix A) does not map in any obvious way onto how the mammalian brain uses self-organization of neurons to extract cognitively significant information from sensory inputs. The Helmholtz machine is a promising alternative, with the added advantage of being able to generate probabilities for alternative explanations for sensory data without external supervision. The internal degrees of freedom of both the recognition and model generation networks consist of layers of "Ising spins" whose activation levels $\{s_i\}$ are either 0 or 1. The Ising spins in the lowest layer are used to represent input data, while the probabilistic activation levels of spins in the "hidden" layers represent the "explanations" for the input data. The probabilistic activation of the Ising spins in the final "output" layer (cf. Fig. 3.1) represents the explanation. The information cost for describing an explanation (α) with this spin ensemble in either the recognition or generative networks is

$$C(\{s_j^l\}) = \sum_l \sum_j (-s_j^l(\alpha)\log p_j^l(\alpha) - (1 - s_j^l(\alpha))\log(1 - p_j^l(\alpha)),$$

$$(3.24)$$

where the $s_j^l(\alpha)$ are the spin excitations in the l'th layer. Because the excitations are stochastic variables, running the recognition network many times over generates a probability distribution Q_α. According to Mumford–Rissanen [26], the best explanation minimizes the total cost $C(\alpha)$:

$$C(\alpha) = -\sum_{\alpha} [Q_\alpha E_\alpha + Q_\alpha \log Q_\alpha - Q_\alpha \log(P_\alpha/Q_\alpha)], \qquad (3.25)$$

where the probability distribution P_α is inferred from the recognition network and the probability distribution Q_α is inferred from the data generation network. The objective of the "wake–sleep" algorithm [9] is to self-consistently lower the free energy, Eq. (1.2), estimated from Bayes's formula to construct MDL representations of sensor data. During the "wake phase", the connection strengths of the top-down model generation network are modified to make the model activation probabilities for the units in the hidden layers align more closely with the actual unit activations that are observed when the bottom-up recognition network is used to interpret examples of input data that are fed to the first layer. This learning step makes the data generation "adversarial" network better at constructing realistic models of the world. In the "sleep phase", the connection strengths in the recognition network are modified so that the activation probabilities for the units in the hidden layers of the recognition network are aligned more closely with the activities of these units that are observed when the top-down network is used to generate "fantasies" of the world. Unfortunately, practical applications of the wake–sleep algorithm have been meager because of the calculations needed to construct the "bottom-up" and "top-down" representations of the input data requiring Monte Carlo sampling of the parameters and variables that cannot be carried out in real time. On the other hand, the introduction of the Helmholtz machine has led to the conceptually important realization that both pattern recognition and RL problems can in principle be approached as the regression problem of reducing the KL divergence between Q_α and P_α, without the need for an explicit model for the system or environment.

The Helmholtz machine also provides another way of viewing optimal control. Indeed, the Bayesian search problem discussed in Section 3.1 can also be regarded as a control problem by interpreting the choice of subsets $\{A_n\}$ as a series of control decisions $U_n \equiv \{u(1), u(2), \dots, u(n)\}$, where $u(t)$ is a map from $\{1, \dots, N\}$ to compact subsets of R^d, which leads to a convergence in a finite time between the choice of search area and the location of the object We observe that these conditional probabilities for control actions can be generated from recursive use of Bayes's formula by assuming that the input training data D consists of N-pairs $(\boldsymbol{x}_n, y_n)$ of d-dimensional vectors $\boldsymbol{x}_n \equiv \{x_1^{(n)}, \ldots, x_d^{(n)}\}$ and scalar labels y_n for each vector. As discussed in the previous section, this training data allows one to immediately construct an interpolating function $z(x)$ for the set of measurements making a prediction for a new measurement result y_{n+1} given an new choice for a compact volume $\boldsymbol{x}_{n+1}$. If Gaussian probability distributions are used for all the probability distributions in Bayes's formula, then Eq. (3.1) provides a probabilistic prediction for the outcome of a control action at step N. A step in the direction of reconciling stochastic estimation based on Bayes's equation and Bellman optimization is to notice that Bayes's formula implies the posterior probability for finding the system in the desired state is proportional to the product of forward and backward filtering probability densities:

$$p_n\left(x = x^* | Y_n, U_{n-1}, p_{n-1}\right)$$
$$\propto p\left(Y_n | U_n, X^* = x, p_{n-1}\right) p_{n-1}\left(x^* = x | Y_{n-1}, U_{n-1}\right).$$

In this way, we recover Todorov's formula, Eq. (1.3), for the *Evidence* factor. Of course, in both quantum mechanics and Todorov's formulation of control theory, the devil lies in the difficulty of taking into account many alternative paths in a path integral representation for the Bellman function.

Chapter 4

Control Theory

4.1. The Hamilton–Jacobi–Bellman Equation

When the state variables are supplemented by control variables, then Bellman's dynamic programming condition, Eq. (2.10), becomes the Hamilton–Jacobi–Bellman (HJB) equation [21,87]:

$$\frac{\partial V^*(X,t)}{\partial t} = \min_{u(t)} \left[l(x,u,t) + \dot{x}(t,x)\frac{\partial V^*(X,t)}{\partial x} \right], \qquad (4.1)$$

so named because of its similarity to the classical Hamilton–Jacobi equation [88]. In (4.1), V^* is the optimum value of the Bellman cost function, and $l(x,t,u)$ is the notation we shall use from now on for the rate of change of the Bellman cost function along a controlled path of the system in state space. In the case of "linear-quadratic" control theory [19], $l(x,u) = q(x) + \frac{1}{2}u^T Ru$, and Eq. (4.1) becomes

$$\frac{\partial V^*(X,t)}{\partial t} = \min_{u(t)} \left[u^T Ru + x^T Q\dot{x} + x(t,x)\frac{\partial V^*(X,t)}{\partial x} \right], \qquad (4.2)$$

If one assumes that the Bellman function is a quadratic form $x^T Sx$, and that Kalman's linear dynamical law $\dot{x} = Ax + Bu$ describes the passive plus controlled dynamics, the HJB equation becomes

$$x^T \dot{S}x = -\min[u^T Ru + x^T Qx + 2x^T SAx + 2x^T SBu], \qquad (4.3)$$

where the capital letters stand for time-dependent matrices acting on the (x,u) vector space. If we minimize the r.h.s with respect to

u, we obtain the optimum control

$$u^* = -R^{-1}B^T S x(t). \tag{4.4}$$

This formula for the control variable is useful in many circumstances of practical interest, which sensibly depends on the deviation of the state variables from the passive dynamics. This noise variance R is a result of an intrinsic noise in the system dynamics, and as we shall see in what follows, increasing this noise is an easy (and often quite effective) way for an adversary to thwart successful control of a system. Equation (4.2) can also be written as a nonlinear ordinary differential matrix equation

$$S'(t) = -SA - A^T S + SBR^{-1}B^T S - Q. \tag{4.5}$$

The ability to rephrase the Bellman recursion relation for optimal control as an ordinary differential equation, which can then be solved with pedestrian numerical integration techniques, is one of the reasons the Kalman filter was so successful.

4.2. Pontryagin Maximum Principle

The solutions of optimal control problems are often represented as streamlines in $(x, \dot{x})$ space. In principle, these streamlines can be found by solving the HJB equation, Eq. (4.1), for various initial values x_0. This equation is similar to the classical Hamilton–Jacobi equation except that the velocity $\dot{x}$ is identified as a control variable u, and the Hamiltonian function that appears in the HBJ equation (the "control Hamiltonian") is minimized w.r.t to u. Just looking at the minimization of the control Hamiltonian is often sufficient [86,87] to find an adequate solution to the control problem.

The deterministic limit of Bayesian feedback control or RL is encapsulated in Pontryagin Maximum Principle, and asserts that the optimum control path is determined by the requirement that the control Hamiltonian in Eq. (4.1) vanishes identically along the control path. In fact, combing the vanishing of the r.h.s of Eq. (4.6) with the Eq. (4.13) for $\dot{\lambda}$ and replacing $u^*(t)$ with $\dot{x}^*(t)$ does lead to the usual classical Euler–Lagrange equations, whose solution is the classical path $x^*(t)$ for which the value of the action functional

$S[x(t)]$ is stationary w.r.t variations in the path. Although the form of the classical equations for x and p is preserved when the position and momentum become quantum operators, the passage from quantum mechanics to classical mechanics by minimizing the phase of the Feynman path integral involves a subtlety that doesn't appear in the textbook treatments of obtaining classical mechanics by minimizing the Maupertuis action functional [88]. In particular, the usual optimization procedure certainly ensures that as Planck's constant go to zero, all the trajectories $x(t)$ in some continuous neighborhood of the optimum trajectory are close to the optimum trajectory $x^*(t)$. However, this doesn't ensure that the time-dependence of the classical momentum $p^*(t)$ emerges in an obvious way. (As noted in the introduction, this greatly puzzled the founders of quantum mechanics.)

The opposite limit of allowing wide deviations in the state of a system from the optimal path is to assume that the only paths which contribute to Eq. (4.9) are paths near the optimal path. Indeed, just as is often the case in quantum mechanics that only paths near the classical path contribute, and in a similar way it is often the case that the only paths that are important in the sum in Eq. (4.9) are near the optimal path. This path is determined by the Pontryagin Maximum Principle [86,87], which is the control theory analog of the Moupertuis principle of least action in classical mechanics. Pontryagin principle gives the fundamental necessary condition for a controlled trajectory to be optimal in the sense of optimizing the Bellman value function.

A fundamental equation for Bayesian learning is the Bellman equation, which at least in principle allows one to solve optimal control problems by minimizing the integral of the rate of change $l(x, u)$ of the Bellman value function over all possible paths. This equation is similar in form to the classical Hamilton–Jacobi equation that provides a geometric optics-like formulation for classical mechanics:

$$\frac{\partial S}{\partial t} = -L(x, \dot{x}, t) - \frac{\partial S}{\partial x}\dot{x}(x, t), \qquad (4.6)$$

where $S(x, t)$ is the classical action function (cf. [83]). Equation (4.6) describes the time dependence of $S(x, t)$ along a path $x(t)$. This path is determined by minimizing $S[x(t)]$ in accordance with Maupertuis's principle of least action [88]. The name geometric optics refers to the fact that in optics path $x(t)$ of a light ray can also be described as the normal to a surface of constant phase for a solution of the wave equation. The limit of geometric optics for say light waves is where the wave nature of light is neglected, and the light is assumed to propagate along a one-dimensional curve determined by the index of refraction for the medium it is propagating through. In classical mechanics, the index of refraction is replaced by a potential, and the path is determined by the condition that the function $L(x, \dot{x}, t)$ (known in classical mechanics as the Lagrangian function) along the path is stationary with respect to variations in the path. The function $L(x, \dot{x}, t)$ also plays an important role in quantum mechanics (cf. [33]). Indeed the Hamilton–Jacobi equation can also be written in a way that is familiar from elementary quantum mechanics:

$$\frac{\partial S}{\partial t} = -H\left(t, x, -\frac{\partial S}{\partial X}\right), \tag{4.7}$$

where the Hamiltonian function for a particle depends on the position x and momentum p of the particle. In writing Eq. (4.7), we have used the Dirac equation $p = -i\frac{\partial S}{\partial X}$ to eliminate the dependence on p. Thus, Eq. (4.7) is a nonlinear equation for classical path. In a similar way in optimal control theory there is an equation for the Bellman value function, $V^*(t, x^*, u^*)$; the HJB equation, which describes the time dependence of the value function along an optimal control path (x^*, u^*)

$$\frac{\partial V^*}{\partial t} = l(x^*, u^*, t) - \frac{\partial V^*}{\partial x}f(x, u^*, t), \tag{4.8}$$

where f plays the role of velocity, and $l(x^*, u^*, t)$ is the sum of a non-stochastic "reward" function that describes how likely the control action leads to the desired state, and a stochastic KL divergence term. We see from Eq. (4.8) that the rate of change of the Bellmann cost function is analogous to the Lagrangian function classical mechanics. Actually, Pontryagin control theory is mathematically equivalent to classical mechanics, and in some respects it is a better formalism than the usual Euler–Lagrangian formalism one finds in textbook accounts of classical mechanics (see e.g. [88]).

An important difference is that the Euler–Lagrange equations only yield a "weak optimization" for classical paths where $x^*(t)$ is the optimum trajectory among trajectories $x(t)$ within some continuous neighborhood of the optimum trajectory. On the other hand, the Pontryagin Maximum Principle yields a "strong" optimization $x^*(t)$ and $u^*(t)$ where the state variable $x(t)$ and control variable $u(t)$ are simultaneously within a continuous neighborhood for both $x^*(t)$ and $\dot{u}^*(t)$. This strong minimum can be found by first identifying the control variable $u(t)$ with a velocity field $\psi(x,t) = \dot{x}$ by an equation of the form

$$\psi(x,t) = f(t,x,u). \tag{4.9}$$

For example, in linear-quadratic regulator theory (cf. Ref. [11]), one writes

$$\psi(x,t) = Ax(t) + Bu(t). \tag{4.10}$$

The optimum velocity field can then be found geometrically [82] by minimizing the Weierstrass function

$$E(t,x,\psi,y) = l(t,x,y) - l(t,x,\psi) - \frac{\partial l}{\partial x}(t,x,\psi)(y - \psi), \tag{4.11}$$

which is a convex function of y. This is equivalent to minimizing the control Hamiltonian

$$H_C(x,u,\lambda,t) = l(x,u,t) + \lambda f(x,u,t), \tag{4.12}$$

where the parameter λ plays much the same role as the momentum in classical mechanics. The variables x and λ satisfy the usual Hamilton equations w.r.t the H_C. For example, if (x_*, u_*) is an optimal trajectory, the equation of motion of λ is

$$\dot{\lambda} = -\frac{\partial L}{\partial x} - \lambda\frac{\partial f}{\partial x}, \tag{4.13}$$

where the derivatives are evaluated at (x_*, u_*).

In linear-quadratic regulator (LQR) theory [21], the reward function is approximated by a quadratic function in x and/or u. For example, a model Hamiltonian might be

$$H_C(x,u,\lambda,t) = \frac{1}{2}[(u^T Ru) + (x^T Qx)] + \lambda(Ax(t) + Bu(t)). \tag{4.14}$$

At its minimum value, H_c is insensitive to u

$$\frac{\eth H_C}{\eth u} = u^{*T}R(t) + \lambda B(t) = 0 \tag{4.15}$$

This "classical" optimum value for the time history of control variable $u^*(t)$ leads to the insight that at least for linear-quadratic regulator theory the control variable is essentially a "momentum" variable

$$u^*(t) = -R^{-1}B^{-T}\lambda^T. \tag{4.16}$$

Of course, this is a conundrum for quantum mechanics in that in quantum mechanics the position and momentum of a particle are treated differently. We will address this enigma in Chapter 7.

4.2.1. *The Moon Lander problem*

The moon lander problem [87] is a simple example of how Pontryagin Maximum Principle works in practice, and is characterized by three simple equations:

$$\dot{h} = v, \quad \dot{v} = -\dot{g} + \frac{u}{m}, \quad \dot{m} = -ku, \tag{4.17}$$

where h is the lander's height above the surface, v is the lander's vertical velocity, m is the lander mass (including fuel), u is the control variable, and g is the acceleration of gravity. The problem resembles the SO(3) Lie–Poisson problem, in that there are three momentum variables; the Lagrange multipliers for v, and $-g + u/m$, and ku, respectively. The control Hamiltonian consists of a piece proportional to u, and the $\Sigma p_i x_i$ piece. It is inherent in this problem that the optimum control is for $u^* = 0$ until a certain time, and then $u^* = 1$ until the landing is complete. The time when one should kick up the thrust to $u^* = 1$ can be determined by back-tracking the solutions to Eq. (4.17) with $u^* = 1$ from $h = 0$ and $v = 0$ at $t = T$ to the point where this trajectory in $\{h, v\}$ space intersects Galileo's parabolic trajectory for a falling body starting at some $h = h_0$ and $v = v_0$. Of course, the back-tracking can only be continued for a time = fuel mass/k:

The state of the system is a two-dimensional vector $x = (h, v)$ with equation of motion:

$$\dot{x} = \begin{pmatrix} 0 & 1 \\ -1 & 0 \end{pmatrix} x + \begin{pmatrix} 0 \\ 1 \end{pmatrix} u, \tag{4.18}$$

while the Hamiltonian having the form

$$H = \lambda_0 + \lambda \left[\begin{pmatrix} 0 & 1 \\ -1 & 0 \end{pmatrix} x + \begin{pmatrix} 0 \\ 1 \end{pmatrix} u \right]. \qquad (4.19)$$

The equations of motion for the Lagrange multipliers $\lambda = (\lambda_1, \lambda_2)$ that follow from Hamilton's equation for "momenta" are

$$\dot{\lambda} = -\lambda \begin{pmatrix} 0 & 1 \\ -1 & 0 \end{pmatrix} \qquad (4.20)$$

The complete solution to the moon landing problem involves [87] matching the trajectory for free fall with the controlled trajectory integrated backward from the moment the lander came to rest.

4.3. Lie–Poisson Dynamics

Typically, "Lie–Poisson" dynamics [89] is defined by an evolution equation of the form

$$\dot{g} = f(A + u_1 E_1 + \cdots + u_l E_l), \quad f \in TG, \ u \in \mathbb{R}^l, \qquad (4.21)$$

where I and the Es are elements of the Lie algebra $\mathbf{g}$ for G (A is called the "drift" term). A controlled trajectory is a pair $(g(t), \mathbf{u}(t))$ where $g \in G$ specifies the "position" within the group manifold, $\mathbf{u}$ is a control vector, and $g(0) = g_0$ and $g(T) = g_T$. Following Bellman's description of optimal control, the optimal control problem is defined by optimizing the integral of a cost rate function $l(u)$ over all possible controlled trajectories $(g(t), \mathbf{u}(t))$ for $0 < t < T$:

$$\int_0^T l[g(t), \mathbf{u}(t)] dt \to \text{minimum}. \qquad (4.22)$$

Typically, $l(\mathbf{u})$ is chosen to be a quadratic function of the control variables

$$l(u) = \frac{1}{2}(c_1 u_1^2 + \cdots + c_l u_l^2), \qquad (4.23)$$

It can be shown that finding the controlled paths which optimize the integral of this function can be found by minimizing the control

Hamiltonian H_C, which differs from the classical Hamiltonian only in that the sign of $l(u)$ reverses the sign of the classical *Lagrangian*, so that

$$H_C(u, q) = l(u, q) + p\dot{g}. \qquad (4.24)$$

In a similar way to how the Euler–Lagrange equations follow from the classical principle of least action, it can be shown that the Lie–Poisson variational principle yields a Lie group analog of Hamilton's equations:

$$\dot{g} = g\, dH_\alpha, \quad \dot{p} = ad^*(dH_\alpha). \qquad (4.25)$$

In general, the dynamics for Lie–Poisson systems can be found [89] by numerically integrating Hamilton's equations along the fibers in T*G with coordinates $(\mu - \beta_0)$, where both the fiber coordinate μ and the constant β_0 are elements of the dual Lie algebra g^* for the symmetry group G underlying the integrable dynamics.

4.3.1. *Rigid body attitude control*

A simple control problem that is important in astronautics and illustrates how the Lie–Poisson dynamics works is provided by the problem of controlling the attitude of a 3D rigid body [3]. The classical Hamiltonian for a freely rotating rigid body in three-dimensions has the form

$$H = \frac{1}{2}\left(\frac{J_1^2}{A_1} + \frac{J_2^2}{A_2} + \frac{J_3^2}{A_3}\right), \qquad (4.26)$$

where the Js are the components of the angular momenta with respect to the axes where the moment of inertia tensor is diagonal, which can also be identified as the Lie algebra generators for SO(3) rotations. The coefficients of this quadratic form define the moment of inertia tensor. Hamilton's equations imply that the time derivative of the angle variable conjugate to the J_i that is a constant of motion is just the constant J_i/A_i. As was shown by Poisson in connection with his investigations of the rotation of astronomical bodies, conservation of energy and angular momentum allows one to reduce the six-dimensional phase space for a "weightless" rigid body floating in space to a two-dimensional phase space, the Hamiltonian becomes:

$$\frac{1}{2}\left(\frac{J_1^2}{A_1} + \frac{J_2^2}{A_2} + \frac{J_3^2}{A_3}\right) \rightarrow \frac{1}{2}\left(\frac{\sin^2\phi}{A_1} + \frac{\cos^2\phi}{A_2}\right)(J^2 - L^2) + \frac{L^2}{2A_3}.$$

$$(4.27)$$

The reduced phase space here is just the 2-sphere $= SO(3)/SO(2)$ with latitude and longitude coordinates (L, ϕ), and the classical motion is just uniform rotation about an axis that is fixed in space. The quantum problem for rotation of a rigid body is a bit more complicated, and involves solving a matrix equation for $2J + 1$ quantum energy levels. The problem of controlling the attitude of a rigid body involves the entire six-dimensional phase space. However, if we have a quadratic control function of form (4.26), the dynamics remains reducible under three-dimensional rotations and corresponds to the intersection of the sphere $J =$ constant and the ellipsoid $H_c =$ constant. It turns out that the exact control dynamics can be expressed in terms of Jacobi elliptic functions. This is our first example of a special analytic function providing an exact solution for a controlled system.

4.4. H_∞ Control

It has been recognized for some time that there is an extension of the Kalman filter, H_∞ control [84], that can be interpreted as a 2-person game. The Kalman filter approach to optimal control has some serious shortcomings: the Kalman filter seeks to minimize the error $\tilde{z}(t) = (z(t) - \hat{z}(t))$ in a series of estimates $\{x_k\}$ for the state of the system based on a series of signals $\{z_k\}$ based on a series of observations $\{y_{k-}\}$ of. Unfortunately, the Kalman control method does not prevent the estimated state of the system from deviating very significantly from the actual state. One possible reason for a large deviation is that an "adversary" can increase the level of noise in the innovations $(z(t) - \hat{z}(t))$. This would directly increase the uncertainty in the state of the system and potentially drastically impact the effectiveness of control actions. This possibility shows up in real-life situations such as the drama in Fig. 1.1. It has also been recognized for some time that there is an extension of the Kalman filter, H_∞ control [84], that can be interpreted as a two-person game. In particular, H_∞ control generalizes the Kalman filter by introducing a limit on the magnitude of the variance of the innovation GP for the Kalman filter. The "H" here stands for the Hardy spaces of holomorphic functions introduced by Segal and Wilson [61] in connection with finding exact solutions for the KdV equation. Let us rewrite the Kalman filter equations in the form

$$x_{k+1} = F_k x_k + w_k, \ y_k = H_k x_k + v_k, \qquad (4.28)$$

where v_k is the measurement noise and w_k is the innovation noise. The goal of the adversary in the H_∞ game is to maximize the uncertainty in the innovation $z_k - \hat{z}_k$, where these zs are combinations of the xs that appear in Eq. (4.28) and are convenient for characterizing the state of a system or environment. The goal of the observer/controller is to minimize by his actions the largest possible estimation error that can occur because of an adversary's actions to increase the innovation noise w_k. The solution to this problem is found by considering how the sum J of all the signal variances normalized by the sum Σ of all the RMS noise variances (including the RMS uncertainty in the initial state x_0) depends on the filter parameters, where

$$J = \frac{\sum_{k=0}^{k=N-1} |Z_k - \hat{Z}_k|^2}{\Sigma}. \qquad (4.29)$$

The minimax procedure [84] for limiting the estimation error is to first maximize J w.r.t x_0 and w_k, assuming that x_k and y_k satisfy Eq. (4.28), and then finding a stationary point of J w.r.t to x_k and y_k. It can be shown that maximizing J w.r.t x_0 and w_k is equivalent to finding the minimum of a control Hamiltonian

$$\mathcal{H}_C(k) = \mathcal{L}_k + \frac{2\lambda_k^T}{\Delta}(F_k x_k + w_k), \qquad (4.30)$$

where $\mathcal{L}_k = |x_x - x|^2 - J_{\max}(|w_k|^2 + |y_k - H_k x_k|^2)$ plays the same role as the classical Lagrangian. This leads to the Pontryagin equations of motion:

$$\frac{\partial \mathcal{H}_k}{\partial \lambda_k} = 0, \quad \frac{2\lambda_k^T}{\Delta} = \frac{\partial \mathcal{H}_k}{\partial x_k}, \qquad (4.31)$$

which are the control theory analogs [86,87] of Hamilton's canonical equations of motion for classical mechanics. The Pontryagin solution for the state variables is an affine Lie–Poisson flow):

$$x_k = \mu_k + P_k \lambda_k^T \qquad (4.32)$$

for the state variables. The solution to Eq. (4.31) that minimizes $J_{\max}$ w.r.t to x_k and y_k is $\hat{x}_k = \mu_k$ and $y_k = H_k \mu_k$. Evidently, given the existence of a bound on the magnitude of J, we obtain a quiescent equilibrium state. Thus, H_∞ control does seem to provide relief for an adversarial increase in the innovation noise.

In contrast with the Kalman filter, the variance of the innovation is bounded. In other words, the actual forward (or backward) phase space trajectories for a system or environment are uniformly close to the observed trajectories. (This is also the miracle of Pontryagin control [86,87]). The geometric and topological proximity of these trajectories allows one to picture the forward and backward "innovations", i.e. two fluctuating smooth surfaces representing differences between the model and observed trajectories viewed from the perspective of the observer/controller or environment acting as the RL agent while the degrees of freedom of the adversary are frozen. The areas of these two surfaces are just the Bellman values V of the actions of the two agents. In the von Neumann equilibrium state, the two surfaces have the same shape, but the Bellman values have opposite signs due to the reversal of the direction of time. (In quantum mechanics, systems propagating backward in time have negative energy [64].)

Chapter 5

Integrable Systems

5.1. RH Solution of the Airy Equation

It is a little-advertised fact that accurate approximate solutions of the time-independent Schrodinger equation for *any potential* can be obtained by making use of "modified Airy functions" (MAFS) [70]. The Airy functions Ai and Bi are defined as exact solutions of the Airy equation which describes quantum motion of a particle in a linear potential. (See also the NBS special functions handbook)

$$u_{xx} = xu. \tag{5.1}$$

At the same time, the MAFs can be used to construct approximate solutions to the inhomogeneous 1D Helmholtz equation:

$$\frac{d^2\psi}{dx^2} + \Gamma^2(x)\psi(x) = 0, \quad \Gamma^2(x) = \frac{2m}{\hbar^2}[E - v(x)] \tag{5.2}$$

of the form

$$\psi_{\text{MAF}}(x) \equiv C_1 \frac{Ai(\xi)}{\sqrt{\xi'(x)}} + C_2 \frac{Bi(\xi)}{\sqrt{\xi'(x)}}, \tag{5.3}$$

where the Ai and Bi are the two kinds of Airy functions. In an oscillatory regime $\Gamma^2(x) > 0$, the argument of the MAFs is $\xi(x) = -(\frac{3}{2}\int_x^{x_0} \Gamma(x)dx)^{2/3}$, where x_o is a classical turning point. Equation (5.2) provides a very good approximation to the exact solution to Eq. (5.2) in an oscillatory regime for any continuous potential

$v(x)$ (even very near to a turning point where the WKB approximation fails) [70,71].

Airy functions were originally introduced into quantum mechanics
in order to solve the problem of quantum motion in a linear potential
[14], but later made an appearance in connection with the general
problem of relating the oscillating and exponentially decaying solutions of the 1D Schrodinger equation at classical turning points [71].
What is of paramount interest for us is that the solutions (5.3) provide approximate solutions for the time-independent 1D Schrodinger
equation that are accurate for all values of x, and for *any potential*. Thus, the MAFs may be especially useful for analytically representing the progress of feedback control or RL, both of the past
and future. This property of MAFs is shared with solutions of the
KdV and NLS equations and reflects a universal property of functions that satisfy integrable PDEs. They also share the fact that they
have meromorphic integral representations as Cauchy integrals [68].
This analytic behavior and its attendant connection with solvability
is fundamental to our approach to optimal control.

In Landau's Appendix for *Quantum Mechanics* [15] (which has a
colorful history going back to the time he was a postdoctoral fellow in
Copenhagen), he points out that the solutions to the Airy equation
can also be expressed as an integral along a line in the complex plane.
When the contour is the imaginary axis, the representation for the
forward propagating solution is

$$Ai(x) = \frac{1}{\sqrt{\pi}} \int_0^\infty \cos\left(sx + \frac{s^3}{3}\right) ds. \qquad (5.4)$$

A similar integral expression exists for the backward propagating
eigenstate $Bi(x)$. Arnold Its [68] pointed out that these integral representations for the Airy function can also be reformulated as a
Riemann–Hilbert problem (see Appendix B for an introduction to
the Riemann–Hilbert problem, which goes back to Riemann's PhD
thesis). The Riemann–Hilbert problem [67–69,73] is to reconstruct
a function that is analytic in the complement of a closed contour Γ
from the discontinuity of the function along the contour. Applying
this to the case where the holomorphic functions are the two simple
momentum space eigenstates of the 1D Schrodinger equation [13] for
a linear potential yields the representation in Eq. (5.4). Except for
arcs at infinity, the RH contour Γ consists of the real line plus the
60° and 120° diagonal lines in the complex plane. Remarkably, these

same lines played a central role in the "Eightfold Way" scheme for constructing representations of SU(3) [118] that played a historical role in the early understanding of elementary particles with strong interactions. The Riemann–Hilbert problem of interest to us consists in recovering the 2-component analytic function $\Phi(z)$ and the solution of the nonlinear Airy equation from a jump condition across Γ:

$$\Phi^+(z) = \Phi^+(z)G(z), \tag{5.5}$$

where $G(z) = \begin{pmatrix} 1 & \Gamma_k \int \frac{\exp -i(2xs + \frac{8s^3}{3})}{s-z} ds \\ 0 & 1 \end{pmatrix}$ and Φ^+ and Φ^- are the holo-

morphic pieces of Φ defined on the two sides of the RH Γ. The integral in Eq. (5.4) is recovered from the component of Φ corresponding to the $Ai(x)$ solution of the Airy equation with a wave incident from the left:

$$u(x,t) = 2i \frac{\lim}{z \to \infty} z\Phi(z)_{12}. \tag{5.6}$$

Its [68] generalized this construction for the $Ai(x)$ to a solution that include both of the two independent solutions of the Airy equation. The jump contour Γ now consists of the real line (the "I spin" axis) plus the entire "U-spin" and "We spin" axes. The jump condition is modified so that along the added pieces of the "U-spin" and "We spin" axes, the jump matrix $G(z)$ in Eq. (5.4) is replaced by its inverse [68–69]. describe a version of this construction where the analytic matrices $\Phi(z)$ involve MAFs near the turning points. They enjoy nice analytic behavior when $\xi(x)$ is extended to the entire complex plane — with the exception of the origin — which induced T–O to modify the jump contour by adding a circle around the origin where Φ is a 2×2 matrix that describes the two independent MAFs that appear inside and outside this circle around the origin in the complex plane. The matrix Φ is holomorphic inside and outside the jump contour, which anticipates the Segal–Wilson solution for the KdV equation.

The matrix Φ has a Cauchy integral representation in terms of its discontinuity across the jump contour

$$\Phi(z) = \frac{1}{2\pi i} \oint \frac{\Phi^+ - \Phi^-}{s - z} ds, \tag{5.7}$$

which can also be written as

$$\Phi^+(s) = \Phi^-(s)G(s),$$

Landau's integral representation for $\int g(\lambda)d\lambda$ for the Airy function can be recovered by taking the limit $\lambda \to \infty$ of λG_{12}. When the contour is the real axis, the jump condition is

$$\Phi^+ = \begin{pmatrix} 1 - |r|^2 & -re^{-2ixs} \\ re^{2ixs} & 1 \end{pmatrix} \Phi^-. \tag{5.8}$$

In this formalism, the input data which allows one to extract $u(x,t)$ are the reflection coefficients $r(s)$. If the contour is a polygon with $G(\lambda) = M_1M_2, \ldots, M_k$, a product of piecewise constant matrices, then the meromorphic scattering amplitude $S(\lambda)$ can be constructed as a product of τ-functions [68]:

$$\tau_k(Y) = Y(\lambda)M_k, \tag{5.9}$$

where

$$Y(\lambda) = \begin{pmatrix} 1 & \int \frac{g(\mu)}{\mu - \lambda} d\mu \\ 0 & 1 \end{pmatrix}.$$

5.2. The KdV Equation

The event that eventually led to the recognition of the remarkable capability of exact solutions of completely integrable PDEs to predict the distant future was the discovery in 1834 by a Scottish marine engineer that the waves created in a shallow canal by a sudden change in motion of a barge can propagate for many miles without changing their shape or speed (the initial observations were limited by the distance the engineer's horse could gallop). Remarkably, it took more than a century for these waves to be understood. (For a history of the efforts to obtain analytic solutions for the KdV equation, see Newell's *Solitons in Mathematics and Physics* [75].) After a prolonged debate within the mathematical physics community as to whether the marine engineer's observations could be believed, a mathematical description of these waves was finally achieved at the end of the 19th century by Korteweg and DeVries [75]:

$$\frac{\partial u}{\partial t} + \frac{\partial^3 u}{\partial x^3} + 6u\frac{\partial u}{\partial x} = 0, \tag{5.10}$$

where $u(x,t)$ is the wave amplitude. Equation (5.14) has solitary wave solutions which are in agreement with the marine engineer's observations. These waves can even undergo collisions with other solitary waves without changing their shapes. It took 60 years after the KdV equation was first written down to find a way of extracting the analytic solutions of the KdV equation which represent solitary waves, and begin to understand why the KdV equation is solvable at all. The first exact solution for the KdV equation was found as a result of the realization by Peter Lax [74] that nonlinear integrable PDEs like the KdV equation can be replaced with an infinite set of linear evolution equations describing the separate evolution in x and an infinite set of times $\{t_n\}$. The solution to these equations is for historical reasons known as the "Baker function" $\Psi(x, \{t_n\})$. This nomenclature goes back to the 1920s [77] and refers to the fact that the eigenvalues of two partial differential operators describing linear evolution in x and t which commute at some time are independent of time, and therefore for each pair of eigenvalues the equations define a time-independent Riemann surface. The Riemann surfaces that are of interest in connection with Bayesian learning are "hyper-elliptic" curves K_n, whose surfaces are parameterized by two variables y and z related by a simple algebraic equation of the form (see Appendix C for an introduction to Riemann surfaces).

The operator L describing the spatial evolution of the Baker function in x is called the Lax operator:

$$L\Psi(x,t) = \lambda\Psi(x,t). \tag{5.11}$$

where we have used t to mean the set $\{t_n\}$. In addition there is a set operators B_n describing the evolution w.r.t the an infinite set of times t_{2n+1}, where $n = 2j + 1$ and t_3 is ordinary time:

$$\frac{\partial \Psi(x,t)}{\partial t_n} = B_n \Psi(x,t). \tag{5.12}$$

As explained in Ref. [77], it is worthwhile to focus on a combination Q of the B_n operators which commutes with L:

$$Q\Psi(x,t) = \mu\Psi(x,t). \tag{5.13}$$

For the integrable differential equations of interest for us the Lax operator L is either the Schrodinger–Hill [125] or a Dirac operator with eigenvalue λ^2 or λ, respectively. The operators $B_n(\lambda)$ are matrix polynomials in the "spectral parameter" λ and $d/d\lambda$. These operators act on a "Baker" wave function $\psi(x, \lambda)$ in a Hilbert space of functions that is generally infinite dimensional, but can also be finite. This nomenclature derives from the coincidence in the KdV case that the Lax operator $L(\lambda)$ is the time-independent Schrodinger–Hill operator with eigenvalue λ^2. The potential for this eigenvalue problem is the physical KdV wave amplitude $u(x,t)$. The eigenfunctions of $L(\lambda)$ are for historical reasons [77] referred to as "Baker functions". For large λ, these functions have the following asymptotic meromorphic form [73]:

$$\psi(x,\lambda) = \left(1 + \frac{a_1(x)}{\lambda} + \frac{a_2(x)}{\lambda^2} + \cdots\right) e^{i\lambda x}. \tag{5.14}$$

The asymptotic behavior of the pre-factor in Eq. (5.5) as $x \to \infty$ is the familiar S-matrix for the 1D non-relativistic Schrodinger equation. If this S-matrix is known, then the exact solutions can be found using the same inverse scattering method originally introduced by Gelfand, Levitan, and Marčenko (see [59,60] and Appendix B) for extracting the potential for the 1D Schrodinger equation from the asymptotic behavior of the pre-factor in Eq. (5.22) interpreted as "scattering data".

It is perhaps noteworthy in this connection that the original noise filter developed by Wiener for extracting a signal from a time series contaminated with white noise also involves using an analytic meromorphic function of frequency (a meromorphic function is a rational function of a complex variable which can have isolated pole singularities) to represent the noise filtered signal. In switching from the real valued spectral parameter λ appearing in the Lax equations to a complex variable ζ, we are acknowledging our fundamental thesis that integrable PDEs represent a promising new approach to the Bayesian model selection problem where conventional regression methods, e.g. Markov chain Monte Carlo methods, fail. This is why machine learning problems can potentially be greatly simplified by transporting the problem to a Riemann surface.

$$y = \pm\sqrt{a_0 + a_1 z + a_2 z^2 + \cdots a_n z^n}. \tag{5.15}$$

In both the KdV and NLS cases, this Riemann surface arises as a corollary of the Burchnall–Chaundry theorem [77] for commuting scalar differential operators. The wave function $\phi(z, x)$, as well as the functions $q(x)$ and $p(x)$, can be calculated exactly in terms of the Θ-functions associated with this Riemann surface [56].

The integrable structure for the KdV and NLS equations is largely hidden. Indeed, initially it was not even suspected that these equations were completely integrable. However, following a decade-long campaign by some talented mathematicians, this hidden structure was finally revealed (see [72–75] for nice reviews). The solution of the KdV equation representing a single solitary wave turned out to have the form [75] $u_1(x, t) = -2\eta^2 \operatorname{sech}^2(\eta(x - x_0 + \eta^3 t + \eta^5 t_5 + \cdots))$, where the appearance of an infinite set of independent time variables reflects the fact that the KdV equation is an example of an integrable dynamical system with an infinite number of degrees of freedom. An important development in the theory of the KdV equation was the discovery by Hiroto [76] that that multi-solitary wave solutions of the KdV equation can be represented in terms of Θ-functions:

$$u(x, t) = 2\frac{\partial^2}{\partial x^2} ln \tau(\theta_1, \ldots, \theta_1), \tag{5.16}$$

where for multiple solitary wave solutions of the KdV equation:

$$\tau(x_1, t_2, \ldots, t_N) = \sum_{n_i, n_j \in Z} \exp\left[i\pi\left(\sum_{i,j=1}^{N} in_i n_j T_{ij} + \sum_{j=1}^{N} 2i\theta_j n_j\right)\right]$$

and

$$\Theta(\{\theta_j\}) \equiv \sum_{n \in zg} \exp\left[i\pi\left(\sum_{ij} n_i T_{ij} n_j + 2\pi i \sum_j \theta_j n_j\right)\right]$$

is the Riemann Θ-function (see [54–57] for the definition and properties of Θ-functions). The $T_{ij} \equiv \oint dw_j(B_i)$ are the "periods" for the associated Riemann surface: (5.15). These periods are obtained by integrating a rational holomorphic differential dw_j around one of the "B" cyclic paths on the surface of the Riemann surface corresponding to going around the circumferences of the "donut holes" in the Riemann surface (see Appendix C). For each donut hole in the Riemann surface, there are two types of cycles corresponding to how one goes around the donut hole: the "A" cycles correspond to a cyclic path threading a donut hole, while the "B" cycles correspond to the cyclic paths going around the circumference of each donut hole.

Given the τ-function in Eq. (5.16), the solution for Eq. (5.11), i.e. the "Baker function", for the KdV equation becomes

$$\psi(x, P)$$

$$= e^{-i\sum_k x_k z_k} \tau\left(x_1 - \frac{1}{z}, x_2 - \frac{1}{2z^2}, x_3 - \frac{1}{3z^3}..\right) / \tau(x_1, \ldots, x_N).$$

$$(5.17)$$

In all these equations, x_1 represents the distance along the "canal", while the other x_is represent an infinite sequence of times $\{t_i\}$ associated with the infinite number of commuting Hamiltonian flows associated with the complete integrability of the KdV equation. It can be shown [75–76] that the expression for $\psi(z, x)$ in Eq. (5.17) is consistent with the expression for $u(x, t)$ in terms of the τ-function given in Eq. (5.16). Also, the appearance of a ratio of τ-functions in Eq. (5.17) explains the emergence of a meromorphic (as opposed to holomorpic) structure for $\psi(x, P)$ since the exact τ-function in both the KdV and NSE cases is identically a Riemann Θ-function, which has zeros but no poles; indeed, the positions of these zeros of τ-functions become parameters in the exact solutions for $\psi(z, x)$ and $u(x, t)$. Apart from these zeros, the initial data needed to define these exact solutions consists of the scattering data for the Baker function as a function of the wave number of the incoming wave number; for example, the reflection coefficient as a function of the wave number. The exact solution for the KdV equation can be found [75] using the inverse scattering method developed in the 1950s by Marčenko, Gelfand, and Levitan (see [61] and Appendix B for more details of the GLM method). Lax showed that the Hamiltonian flows for the Baker function leave the eigenvalues of the $L(u)$ operator in Eq. (5.11) intact, but evolve the potential $u(x,t)$ for the Lax equation in such a way to generate a solution to (5.14).

The asymptotic Baker function can be written in the form:

$$\psi(x, \{t_n\}, \varsigma) \sim \frac{X(\varsigma)\tau}{\tau}, \tag{5.18}$$

where

$$X(\varsigma) = \exp\left(i\sum_0^\infty \varsigma^{2k+1} t_{2k+1}\right) \exp\left(i\sum_0^\infty \frac{\varsigma^{-2k-1}}{2k+1} \frac{\partial}{\partial t_{2k+1}}\right). \tag{5.19}$$

The model selection problem for the KdV equation amounts to choosing a set of solitary waves and values for the initial positions x_{i0} and "momenta" η_i that best explain a set of observations of the wave amplitude that, say for practical reasons, are limited in their scope of times and locations. Finding the best choice of parameters for the τ-function in Eq. (5.16) based on actual video observations of wave amplitudes would be a very difficult problem for conventional machine learning if many solitary waves were present.

In the context of using the KdV equation as an avatar for Bayesian learning, this model selection problem amounts to choosing a "Backlund transformation" [75]. This involves transforming the τ-function to accommodate the addition of another solitary wave. Using the formula for $u(x,t)$ as a second derivative of the τ-function one finds that the new τ-function can be expressed in terms of asymptotic scattering wave functions for the KdV Lax equation:

$$\tilde{\tau} = \tau(A\psi_+(x,\varsigma) + B\psi_-(x,-\varsigma)) = (AX(\varsigma) + BX(\varsigma))\tau, \tag{5.20}$$

where A and B are different constants for each application of the Backlund transformation (as well as independent ς variables for each application of the Backlund transformation since, after all, the

Riemann surfaces after each transformation are quite different!). To get some idea of what is going on here, we note that the τ-function could be written in the following form:

$$\left(\sum_{k=1}^{\infty} b_{rk} x_k z^{-k}\right)\left(1+\sum_i a_i z^{-i}\right)$$

$$= \exp\left\{\sum_i \lambda_i \left(x_i - \frac{1}{z^i}\right) + \sum_{ij} \mu_{ij}\left(x_i - \frac{1}{z^i}\right)\left(x_j - \frac{1}{z^i}\right)\right\}$$

$$\Big/ \exp\left(\sum_i \lambda_i x_i + \sum_{ij} \mu_{ij} x_i x_i\right). \tag{5.21}$$

Thus, if we introduce new terms with negative powers of ζ to the phase of the Baker function, we end up with something that looks like Hiroto's KdV solution with an extra solitary wave. This manifestly changes the topology of the Riemann surface. As we shall see in Chapter 7, we can to some extent avoid using the Backlund transformations if we simply start with a Riemann surface with sufficiently high genus.

5.3.   Segal–Wilson Construction

The work of Its [68] set the data science community on the path connecting the use of the inverse scattering method to solve nonlinear integrable PDEs to construct special meromorphic functions in a neighborhood of the north pole of a Riemann surface, which can be used to construct a 1:1 map between input data to data features. Following the success of the inverse scattering method for constructing exact solutions of the KdV or NSE equation, Segal and Wilson discovered [73] a nice geometric way of side-stepping the usual way of solving the GLM integral equation. In the Segal–Wilson approach, scattering amplitudes for the Schrodinger or Dirac equations appears as a discontinuity between square-integrable holomorphic functions defined in the upper and lower halves of the complex plane. Their construction is based on the introduction of the "Grassmannian" of all closed subspaces W of the Hilbert space H consisting of square integrable functions that are close to the subspace $H+$ spanned by the $\{z^i\}$ where $i \geq 0$. "Close" to $H+$ means that the orthogonal projection $W \to H_+$ is transversal to H_-; i.e. it consists of complex analytic functions of the form

$$w(z) = \sum_{k=-N}^{\infty} a_k z^k. \tag{5.22}$$

The Segal–Wilson approach to solving the KdV equation begins with the introduction of "loop maps":

$$g : S^1 \to H_+ + H_-, \tag{5.23}$$

which map the equator of the Riemann sphere (S^2) into a curve W in the space of complex $n \times n$ matrices representing $H = H_+ + H_-$. In this language the usual least squares regression problem described in Chapter 2 becomes the problem of factorizing the loop map g *a la* Wiener–Hopf (cf. Appendix B) into the product $g^- g^+$ of two maps g^+ and g^- that are holomorphic in the upper and lower hemisphere's S_+ and S_-, respectively, of the Riemann sphere. Taken together H represents a meromorphic extension of the space H_- of holomorphic functions representing input data. In the case of the 1D inhomogeneous wave equation, the spaces H_+ and H_- corresponding to all possible outgoing and incoming states are related by

$$g(\lambda) = \begin{pmatrix} a_\lambda & b_\lambda \\ c_\lambda & d_\lambda \end{pmatrix}. \tag{5.24}$$

The usual scattering matrix S_λ in 1D is defined by

$$g(\lambda) = e^{-\lambda J_z} S_\lambda e^{\lambda J_z}, \tag{5.25}$$

where

$$S_\lambda = \frac{1}{d_\lambda} \begin{pmatrix} 1 & b_\lambda \\ -c_\lambda & 1 \end{pmatrix} \equiv \begin{pmatrix} T_\lambda & R_\lambda \\ -R_\lambda & T_\lambda \end{pmatrix} \tag{5.26}$$

R_λ and T_λ are the usual reflection and transmission functions for a wave packet incident on a localized potential from either the left or right. The factorization problem for $g(\lambda)$ amounts to writing

$g^- = g(g^+)^{-1}$. If we write $g^- = g(g^+)^{-1}$, $g = 1 + \gamma$, $g^+ = 1 + \gamma_+$, and $g^- = 1 + \gamma_-$, then [73]

$$\gamma_- = \gamma + \gamma_+ + \gamma\gamma_+. \tag{5.27}$$

Finding the two factors is reminiscent of the Wiener–Hopf method of representing scattering amplitudes as a ratio of factors that are holomorphic above and below the real axis in k^2 space. However, Eq. (5.27) is a significant improvement over the usual least squares regression method which depends on the inversion of a large matrix to solve the Weiner–Hopf equation. Indeed, Eq. (5.27) doesn't require a matrix inversion, but instead is a product of matrices.

The crucial part of the Segal–Wilson construction is understanding how the matrices representing g^+ and g^- act within H. The part of the map $gH_+ \rightarrow H_-$ connecting H_+ and H_- yields the meromorphic pre-factor $A(x, z)$ in Eq. (5.14) by providing a map $w : H_+ \rightarrow H_-$ whose graph is just this sought after 1:1 connection between the spaces H_+ and H_-. That is gH_+ can be written in the following form:

$$gH_+ = \begin{pmatrix} 1 & 0 \\ w & 1 \end{pmatrix} H_+. \tag{5.28}$$

Because $g^{-1}(gH_+)$ leaves H_+ invariant, we obtain

$$g^{-1} \begin{pmatrix} 1 & 0 \\ w & 1 \end{pmatrix} = \begin{pmatrix} a & b \\ 0 & c \end{pmatrix} \tag{5.29}$$

as the solution of the problem of factoring the map g into the two factors $g+$ and g^- holomorphic on H_+ and H_-, respectively. Overall, the Baker function is a product of an exponential factor and the function $w(z)$. Segal and Wilson are acknowledging that an essential element in understanding the KdV and NLS equations is the replacement of the real valued spectral parameter λ with a point P on a Riemann surface, which is often referred to as the spectral curve. The spectral curves that are of interest in connection with machine learning are "hyper-elliptic" curves K_n, whose surfaces are parameterized by two variables y and z related by the algebraic equation in Eq. (5.15).

The Segal–Wilson provides a geometric construction of the Baker function $w(z)$ on a hyper-elliptic Riemann surface. In both the KdV and NLS cases, the meromorphic pre-factor in the asymptotic Baker function has the general form $\phi_W(x, P) = \left(1 + \sum_{i=1}^{\infty} a_i(x)/z^i\right)$, and can be represented as the ratio of two τ-functions $\tau_W(z)$. For certain discrete combinations (z_k, t_k), the quantity $\exp\left(-\sum_k z_k t_k\right) W$ is transversal to H_-; i.e. it has the form

$$\exp\left(-\sum_k z_k t_k\right) w = 1 + \sum_{i=1}^{\infty} a_i(x)/z^i. \tag{5.30}$$

If the t_k in Eq. (5.30) are nonzero, then the action of the exponential factor in Eq. (5.17) represents the effect of the multiple flows on the solution $u_w(z)$, provided that these are independent linear flows for each value of k, which of course makes sense since the KdV equation is an infinitely integrable system. The transversal condition means that the parameters defining w satisfy certain conditions, which in the "Kyoto school" theory of the KdV equation are met by demanding that the Baker function be derived as the ratio of two τ-functions as in Eq. (5.17), which explains the meromorphic structure of the Baker function a la Wiener filter.

In the multi-solitary wave case, these τ-functions can also be represented as a determinant of propagators for solutions of a Fokker-Planck equation [120]:

$$D(z) = \sum_{l=1}^{\infty} \frac{z^l}{l!} \int_a^b ds_1 \ldots \int_a^b ds_l \begin{vmatrix} K(s_1, s_1) & \cdots & K(s_1, s_l) \\ \vdots & \cdots & \vdots \\ K(s_l, s_1) & \cdots & K(s_l, s_l) \end{vmatrix}, \tag{5.31}$$

where each $K(s_1, s_2)$ can be thought of as a propagator for Gaussians localized at points $s = (x, t)$. The $D(s)$ notation for the τ-function in (5.31) has its origin in scattering theory (see Appendix B). The $K(s_1, s_2)$ can also be interpreted as a backward filter for solutions of the time-independent KdV equation

$$Kf \equiv \int_a^b K(s, s')f(s')ds'.$$

The solution for $u(x, t)$ obtained from (5.31) is the same as the analytic expression involving theta functions for a Riemann surface obtained by Hirota [76]. This provides a simple and beautiful solution of the model selection problem for Bayesian learning.

The initial clue that exact solutions for the KdV equation might be connected quantum mechanics was provided by Einstein [91], who showed the Bohr–Sommerfeld quantization rules are exact for completely integrable systems. As we shall see in Chapter 7, this has the consequence that the high frequency limit of multi-channel quantum mechanics can directly yield the reward function for multi-observer control problems or multi-agent RL problems. A significant role for multi-channel quantum mechanics also shows up in the work of Fadeev *et al.* on the NLS equation [126].

5.4. The NLS Equation

The Lax equation approach for finding exact solutions also works for the nonlinear Schrodinger equation [69,75]. Of particular interest to us is its "complexified form" where the scalar wave amplitude $u(x,t)$ of the KdV equation is replaced by two amplitudes $p(x,t)$ and $q(x,t)$, which play the role of momentum and position controls:

$$i\frac{\partial q}{\partial t} = -q_{xx} + |p|^2 q$$

$$i\frac{\partial p}{\partial t} = p_{xx} - |q|^2 p. \tag{5.32}$$

If we assume $p = \pm\overline{q}$, then we have the real form analogs to the KdV equation:

$$i\frac{\partial u}{\partial t} = -u_{xx} + 2|u|^2 u, \tag{5.33}$$

which is of particular interest because it has found optical fiber applications. The discovery that the NLS equation, like the KdV equation, is completely integrable [69,73,75] led to the realization that the same kind of inverse scattering approach that worked for the KdV equation also works for the NLS equation.

The NSE differs from the KdV equation in that in the KdV case the physical wave amplitude $u(x,t)$ is real, whereas the NSE is more naturally interpreted as a wave equation for a complex amplitude.

The Lax equation for the complexified NSE is a Dirac-like matrix equation

$$M\Psi(x, z) = z\Psi(x, z), \tag{5.34}$$

where

$$M = i \begin{pmatrix} d/dx & -q \\ p & -d/dx \end{pmatrix}. \tag{5.35}$$

The Baker wave function has the form

$$\Psi(z, x) = \psi_1 = \exp\left[\int_{x_0}^x (-iz + q\,(x'))\,\phi(z, x)dx'\right] \tag{5.36}$$

$$\psi_2 = \phi(z, x)\psi_1,$$

where $\phi(z, x)$ is a meromorphic function with asymptotic form $(1 + a_1(x)/z + a_2(x)/z^2 \ldots)$. Just as in the KdV case, the variable z refers to a point on a canonical Riemann surface associated with eigenvalues of the Lax operator. Also, as in the KdV case exact solutions for all the quantities in Eq. (5.36) can be found [59] in terms of Θ-functions, and in addition the reward function $q(x)$ can also be expressed in terms of a τ-function in a fashion analogous to Eq. (5.16). As also in the KdV case the appearance of a ratio of τ-functions in the expression for $\phi(z, x)$ explains the emergence of a meromorphic (as opposed to holomorphic) structure for $\psi(x, P)$. However, in contrast with the KdV equation, the nonlinear Schrodinger equation mixes forward and backward propagating modes in essentially the same way that the Dirac equation connects positive and negative energy states when there is a nonzero potential.

Dirac's original motivation was to develop a relativistic version of the Schrodinger equation [35]. The original version of quantum mechanics allowed nonlocal physical effects that violated the principle of relativity. In his *Theory of Positrons* [64], Feynman provided a clear explanation as to why the Dirac equation cures this difficulty. Formally, the way the Dirac equation solves the problem with causality is to include negative energy modes that propagate backwards in time. This also means though [127] that it is more natural to regard the solution of the Dirac equation as a quantum field than a classical wave function as in Schrodinger's approach to quantum mechanics.

For similar reasons in the case of the NLS equation it is almost necessary from the beginning to recognize that the $\psi(x, P)$ amplitudes are quantum fields.

A "second quantized" version of the NLS equation was introduced by Faddeev *et al.* [125]. Their Quantum Inverse Scattering formalism allows one to express the τ-function and Baker function for the second quantized NLS equation in terms of expectation values for products of creation and annihilation operators for the oscillator array. The Hamiltonian for NLS model introduced by Faddeev & Co is the same as the Hamiltonian for discretized version of a 1D gas of strongly repulsive bosons, with interactions:

$$H = \int dx \left[\partial_x \Psi^\dagger \partial_x \Psi + c \Psi^\dagger \Psi^\dagger \Psi \Psi \right], \tag{5.37}$$

where the bosonic operators Ψ_n, $n = 1, \ldots, M$, satisfying the usual commutation relations for Bose fields [127]:

$$\{\Psi_m, \Psi_n\} = \{\Psi_m^*, \Psi_n^*\} = 0, \quad \{\Psi_m, \Psi_n^*\} = \frac{1}{\Delta} \delta_{mn}. \tag{5.38}$$

Actually, for a finite number of oscillators the Hamiltonian (5.37) can be replaced by a many-body quantum mechanical problem defined by a Hamiltonian

$$H_N = -\sum_{j=1}^{N} \frac{\partial^2}{\partial z_j^2}, \tag{5.39}$$

with a boundary condition

$$\left(\frac{\partial}{\partial z_{j+1}} - \frac{\partial}{\partial z_j} - c \right) \chi_N = 0, \quad \text{for } z_{j+1} = z_j + \epsilon.$$

The energy eigenfunctions have the form

$$|\Psi_N> = \frac{1}{\sqrt{N!}} \int dz^N \chi_N(z|\lambda) \Psi^\dagger(z_1) \cdots \Psi^\dagger(z_N)|0>. \tag{5.40}$$

Apart from a normalization factor χ_N, this solution is the Bethe ansatz:

$$\chi_N \propto \prod_{N>j>k} [sgn(z_j - z_k)] \sum_P (-1)^{|P|} \exp\left\{ i \sum_{n=1}^{N} z_n \lambda_{Pn} \right\}, \tag{5.41}$$

where λ is the spectral parameter and $E_N = \sum_{j=1}^{N}(\lambda_j^2 - \mu_F)$. The momentum eigenvalues associated with excitations of the Bethe vacuum are

$$\exp(iL\lambda_j) = (-1)^{N-1}. \tag{5.42}$$

The quantities that would most naturally play the role of the real valued kernel functions $K(x, y)$ that appear in the theory of the KdV equation are the equal time correlation functions

$$Q(x_1, x_2) = \left\langle \int_{x_1}^{x_2} \langle \Psi^T(y)\Psi(y)\rangle dy \right\rangle, \tag{5.43}$$

where the bracket is evaluated in either the Bethe ground state or some combination of excited Fock states. As shown in [126] the quantum correlation function $Q(x_1, x_2)$ satisfies differential and integral equations similar to those satisfied by the self-reproducing kernels in the classical theory of stochastic estimation. A typical result is that when the background state is a thermal state, then the analog of the familiar covariance matrix $K(x, y)$ used for stochastic estimation is

$$K(\lambda, \mu) = \sqrt{\vartheta(\lambda)}\,\frac{\sin(\lambda - \mu)}{\lambda - \mu}\,\sqrt{\vartheta(\mu)}, \tag{5.44}$$

where $\vartheta(\lambda)$ is the Fermi weight $1/\exp(\lambda^2 - \beta)$. The analog of the conventional scattering problem is $|\lambda| \to \infty$. Because we have analytic expressions for these kernel functions, we can naturally make contact between traditional methods of data analysis and our use of integrable models to make predictions for optimal control or RL strategies.

The τ-function $\tau(\lambda)$ is defined to be the trace of the analog of the S-matrix [125]:

$$\tau(\lambda) = trT(L, 1|\lambda), \tag{5.45}$$

which in turn is a product of the lattice displacement operators $L(n|\lambda)$:

$$T(L, 1/\mu) = L(L/\mu)\dots L(1/\mu).$$

Each $L(n|\mu)$ factor is a discretization of the Lax displacement operator:

$$L(n|\mu) = \begin{pmatrix} 1 + (-1)^n \dfrac{\Delta}{4} - i\mu\Delta/2 & -i\Delta\Psi_n^* \\ i\Delta\Psi_n & 1 + (-1)^n \dfrac{\Delta}{4} + i\mu\Delta/2 \end{pmatrix}, \quad (5.46)$$

where Δ is the "lattice spacing". These 2×2 matrices play much the same role as the classical Lax operator for integrable nonlinear differential equations. We have thus come complete circle in the sense the τ-function had its origin in the mathematics of the Ising model [120].

5.5. Galois Remembered

Of course, the foregoing discussion of integrable models begs the question as to whether the use of the KdV or NLS equations to define the reward function for feedback control or RL is sufficiently general to solve all pattern recognition, feedback control, or RL problems of interest. One possible answer to this question is provided by the Helmholtz machine [7,8], where it is not necessary to define *a priori* the models for a system or environment. Instead, the Helmholtz machine uses the input data itself to define an adversarial network scheme to represent the input data, in much the same way that the Boltzmann machine [1] is used to represent a set of data as a thermal state of spins.

The penultimate fundamental discovery that we would like to highlight is the cornucopia of mathematical results that have flowed from the notes that Evariste Galois scribbled in 1832 on the evening before the duel which took his life. Ten years after Galois's death the scope of his accomplishment eventually came to light when Joseph Liouville announced that upon perusal of Galois's notes he found that Galois had solved in a particularly elegant way the problem of determining when an algebraic equation could be solved by combining the usual arithmetic operations of addition, subtraction, multiplication, and division with the operation of taking the nth root of a rational number. Galois's great achievement was to link this problem to group theory [92]. His method involved translating the problem of solving an algebraic equation into a problem in group theory. Following the publication of Galois's idea, several mathematicians pursued the idea of developing similar group theory approaches for solving differential equations. However, these efforts only achieved results that are decisive for our enterprise in conjunction with efforts to solve the "inverse Galois program" of determining whether a given group is the Galois group of some field extension.

Digging a little deeper one can perhaps glimpse that our quantum approach to Bayesian model selection is the fruit of a marriage between Galois's theory of solvability and Weyl's premonition of an intimate link between group theory and quantum mechanics. In particular, beneath the association of integrable differential equations, the theory functions on a Riemann surface and Bayesian inference, we are led back to a Shakespearean tragedy that occurred in 1832, when Evariste Galois lost his life in a duel, just a day after he had scribbled some notes which are now recognized as one of the citadels of human achievement. The focus of his notes was the problem of solving algebraic equations using arithmetic and square root operations in much the same way one learns in elementary algebra how to solve quadratic equations. Galois observed that there is a correspondence between the "normal" subgroups of the automorphism group of the field of rational numbers extended by the roots of a polynomial equation and the arithmetic operations used to solve the polynomial equation (see [92]). A normal subgroup was defined by Galois as subgroup which leaves the rational numbers used to define the field extensions' invariant. As was first glimpsed by David Hilbert [128], the whole structure of using group theory to solve polynomial equations can be replicated by considering the fields of meromorphic functions on coverings of a Riemann surface as an analog of Galois's field extensions of rational numbers by the square roots of prime numbers.

One begins to get a sense that what really underlies Bayesian machine learning is Galois's theory of the solvability of polynomial equations. One area of unfinished business for mathematics in the 21st century is to extend Galois's insights regarding the solvability of polynomial equations to the integrability of differential equations. It is now understood through [93] that the analog of the Galois fields that play a central role in the solvability of polynomial equations is the field of rational functions on a Riemann surface. This is congruent with our expectation that solvability of the KdV and/or nonlinear Schrodinger equation is what underlies Bayesian machine learning since in either case the construction of a certain rational function on Riemann surface plays a central role in constructing both the exact solution of these equations and the transition amplitude in Eq. (1.4). The important role played by a characteristic rational function of a complex variable actually goes back to the original work of Wiener and Kalman on linear filters [83] for signal analysis or optimal control. Indeed, one might consider our quantum approach to automating Bayesian inference as the fruit of a marriage between Galois's theory of solvability and Weyl's premonition of a close link between group theory and quantum mechanics [44].

The business of machine learning typically amounts to determining the meromorphic prefactor in (5.30). This characterization of machine learning in turn suggests a parallel between our quantum mechanical approach to machine learning and Galois theory. In particular, Eq. (5.30) suggests that the space of meromorphic functions on a Riemann surface can be regarded as an extension of the space of holomorphic functions of a complex variable z in a manner reminiscent of the way that Galois theory provides an elegant characterization of the solvability of algebraic equations in terms of an extension of rational arithmetic operations by nth-root operations (For an entertaining introduction to the connection between Galois theory, the solvability of ordinary differential equations, and meromorphic functions, see Kuga's *Galois' Dream* [93]). In the context of using representations of the quantum commutation relations (5.38) (see also Appendix D) to find the form of the meromorphic function in (5.30), the role of group theory in Galois's original approach to finding the roots of algebraic equations will evidently be played by quantum mechanics. To some extent this reflects Hermann Weyl's spotlight [44] on the importance of group theory for quantum mechanics. An intimate connection between the continuous group SU(N) and quantum mechanics will be one of the main threads of our presentation. One of our ambitions in the following will be to assess the plausibility that this thread will eventually lead to practical methods for using quantum mechanics for solving machine learning problems; although admittedly at this time we can only offer theoretical arguments rather than experimental results to support this expectation. Our theoretical arguments will be based on the intuitive ideas that quantum machine learning can be thought of as a form of Galois theory. We can immediately offer the observation [42] that the representations of SU(N) can be classified using the same permutation groups that play an important role in Galois's theory of algebraic equations. The connection between Galois theory and Riemann surfaces also appears [128] in the problem of Bayesian model selection, which shows that the Bayesian model selection problem and its connection with quantum mechanics has very deep mathematical roots.

Chapter 6

Quantum Tools

6.1. Weyl Remembered

In general quantum mechanics, Hilbert spaces can be defined as finite or infinite dimensional vector spaces that give rise to representations for the Weyl–Heisenberg group; i.e. the continuous 3-dimensional group obtained by exponentiation of the Heisenberg commutation relations [44,94]. (In the mathematics literature, this group is called the Heisenberg group. We have restored Weyl's name because this group was first introduced by Weyl.) The generators for the Weyl–Heisenberg group are a shift operator $S(y)$, and an "automorphic" translation $T(x)$:

$$S(y) = \exp(-iy \cdot \partial/\partial\xi), \quad T(x) = e^{ix \cdot y/2} \exp(ix \cdot \xi), \qquad (6.1)$$

where x, y, and ξ are d-dimensional vectors of continuous real variables. Apart from the factor $e^{ix \cdot y/2}$, the operators $S(y)$ and $T(x)$ are familiar as the building blocks $X(s) = \exp(-is\hat{p})$ and $Z(t) = \exp(it\hat{q})$ for universal quantum computations with quantum states parameterized by continuous variables [12].

What originally attracted the attention of mathematicians to the Weyl–Heisenberg group was the observation that the original Heisenberg commutation relations don't make rigorous mathematical sense in some important cases, e.g. in the case of action/angle variables which are used to represent completely integrable classically dynamical systems [91]. According to the celebrated theorem of Stone and von Neumann [94], all irreducible unitary representations of this group have the following form (see also Munford's Tata Lectures [96]):

$$\pi_\lambda(x, y, t)\varphi(\xi) = e^{i\lambda t}e^{i\lambda(x \cdot \xi + \frac{1}{2}x \cdot y)}\varphi(\xi + y), \qquad (6.2)$$

where φ is a square integrable function of n real variables and x, y, and ξ are real vectors. Thus, the general form of the Weyl–Heisenberg group is a translation by y followed by a multiplication by an exponential factor involving x as a "wave number" and a scalar t. This 3D group can also be realized as upper triangular $n + 2 \times n + 2$ matrices of the form

$$I + \begin{pmatrix} 0 & x & t \\ 0 & 0 & y \\ 0 & 0 & 0 \end{pmatrix}. \qquad (6.3)$$

This group is a "nilpotent", which means that the Lie algebra for the Heisenberg group has the same form as the above matrix minus the identity operator. (Incidentally, this is the mathematically rigorous formulation of the original matrix mechanics of Heisenberg, Born, and Jordan [42]).

Among the representations of the Weyl–Heisenberg group, the representations related to the energy eigenstates of a harmonic oscillator will be of special importance to us. In particular, Bargmann introduced a type of quantum coherent state for 1D quantum oscillators, known as the BFS states [94]. The Bargmann–Segal transform [95] is a map $f(x) \to F(z)$ from square integrable functions on Euclidean space R^d to $\mathbb{C}^d$:

$$F(z) = \frac{1}{\pi^d}\int e^{-\frac{z^2 + 2\sqrt{2}x \cdot z - x^2}{2}} f(x)dx \qquad (6.4)$$

The power of the BS transform (6.4) is that a space of holomorphic function defined on a compact domain can be mapped to a compact space of harmonic oscillators with real valued wave functions. Because of the ubiquitous importance of holomorphic functions for Bayesian learning, this result is potentially of great interest to us.

In 1928, Fock had observed [94,95] that regarded as operators in a Hilbert space of holomorphic functions z and d/dz obey the same commutation relations as the annihilation and creation operators for a quantum harmonic. Following Bargmann's original paper [95], Fock introduced as an alternative to the Fock space of energy eigenvalues the vector space of holomorphic functions defined on some domain U of $\mathbb{C}^N$ whose basis is the set $\{z^n/\sqrt{n!}\}$ where $z \in \mathbb{C}^N$, and the normalization integral includes a factor $\mu(z) = \left(\frac{1}{\pi^d}\right) \exp(-|z|^2)$. Of course, the normalization integral must necessarily include an additional factor to make it finite. The reproducing kernel is

$$K(z, w) = e^{z \cdot \overline{w}}. \tag{6.5}$$

This kernel is reproducing in the sense that any holomorphic function $f(z)$ in U can be written in the form

$$f(z) = \int K(z, w) f(w) \mu(w) dw. \tag{6.6}$$

The presence of the factor $\mu(w)$ in this equation means that the displacement operator $f(z) \rightarrow f(z - a)$ is not unitary; instead one represents displacements with a unitary operator

$$T_a f(z) = e^{-\left(\frac{|z||^2}{2}\right) + z \cdot \overline{\overline{a}}} f(z - a). \tag{6.7}$$

As first noted by Fock, the operators $A = d/dz$ and $A^* = z$ satisfy the canonical commutation relations; i.e.

$$[A, A^*] = 1. \tag{6.8}$$

Also, as a result of these commutation relations, the translation operators (6.7) satisfy a composition law

$$T_a T_b f(z) = e^{-iIm(a \cdot \overline{b})} T_{a+b} f(z). \tag{6.9}$$

These BFS states are defined as superpositions of the coherent states for a quantum harmonic oscillator introduced by Schrodinger of the form

$$|\alpha >= \exp(\alpha a^+ - \alpha^* a)|0 >, \tag{6.10}$$

where a^+ and a are the creation and annihilation operators for a quantum harmonic oscillator. These states are also relevant for the Segal–Wilson construction [75], as well as H_∞ control [84]. These BFS states also turned out to be of great practical importance in quantum optics [129].

Writing $D(\alpha)$ for the operator in Eq. (6.10), one has the composition law

$$D(\alpha + \beta) = D(\alpha)D(\beta)\exp(-iIm[\alpha^*\beta]), \qquad (6.11)$$

The last factor in Eq. (6.11) is a signature for the fact that a^+ and a operators obey the Fock commutation relation (6.8). In the position representation commonly used for the Schrodinger equation, these states have the form

$$|\alpha >= \exp\left[\frac{x^2}{2} - \sqrt{2}\alpha x + \frac{\alpha^2}{2}\right]. \qquad (6.12)$$

These $|\alpha >$ states are not orthogonal, but have an overlap

$$| < \beta|\alpha > |^2 = e^{-|\beta - \alpha|^2}. \qquad (6.13)$$

It is of course interesting that these coherent quantum states overlap in a way that resembles the popular squared exponential kernel that is so useful in data analysis. In many ways, the coherent states defined by Eq. (6.12) are the canonical choice for a representation of the Heisenberg group, and by providing an entre for holomorphic functions, these functions play a central role in our quantum representations for Bayesian learning. These states are also closely related to the radial basis functions $\psi(x, 0) = \exp(-\frac{m\omega}{4h}(x - a)^2)$ that are commonly used in machine learning [6]. The propagator for these states is [47]:

$$\iint_{-\infty}^{\infty} \psi_b^*(x_b)K_o(x_b, T; x_a, 0)\psi_a^*(x_a)dx_bdx_a$$

$$= \exp\left\{-\frac{i\omega T}{2} - \frac{L\omega}{4h}\left(a^2 + b^2 - 2abe^{i\omega T}\right).\right. \qquad (6.14)$$

Starting from a state $\psi(x, 0)$, the state after time t is

$$\psi(x, T) = \exp\left\{-\frac{i\omega T}{2} - \frac{m\omega}{4h}\left(x^2 - 2abxe^{-i\omega T} + a^2\cos(\omega t)\right)e^{-i\omega T}.\right.$$

If, in addition there is a linear coupling to another oscillator, Eq. (6.14) becomes

$$F(b, a) = \exp\left\langle -\frac{i\omega T}{2} - \frac{L\omega}{4h}(a^2 + b^2 - 2abe^{i\omega T})\right.$$

$$\left. + \sqrt{\frac{m\omega}{2h}}(a\beta + b\beta^* e^{-i\omega T} + \cdots)\right\rangle,$$

where $\beta = \frac{1}{\sqrt{2m\hbar\omega}}\int_0^T f(t)dt$ and the dots are a term quadratic in the external force $f(t)$ acting on the oscillator.

Another notable way of relating holomorphic functions and quantum harmonic oscillators involves N-dimensional quantum oscillator states and the Wigner–Fourier transform [96]. In particular, any function $f(z \cdot t)$ can be reconstructed from its Fourier–Wigner transform:

$$\tilde{f}(z) = \sum_{\alpha,\beta} (f, \Phi_{\alpha,\beta})\Phi_{\alpha,\beta}, \tag{6.15}$$

where $f \in L_2(\mathbb{C}^N)$. When α, β are integers and $\Phi_{\alpha,\beta}$ has the form

$$\Phi_{\mu,\nu}(z) = \frac{1}{(2\pi)^{N/2}} \int e^{ix \cdot \xi} \Phi_\mu\left(\xi + \frac{y}{2}\right) \Phi_\nu\left(\xi - \frac{y}{2}\right) d\xi, \tag{6.16}$$

where $x, y, \xi \in \mathbb{R}^N$ and $\mu, \nu \in \mathbb{Z}^N$, and $\Phi_\mu(x)$ is the wave function for an array of quantum oscillators where just 1 energy level per oscillator is occupied:

$$\Phi_n = \prod_{i=1}^{N} H_{n_i}(x_i), \tag{6.17}$$

where H_{ni} is the Hermite function for a single Fock state, $n = \{n_j\} \in \mathbb{Z}^N$ and $x = \{x_j\} \in \mathbb{R}^N$. It can be shown that the set of functions $\{\Phi_n\}$ for all n provides a basis for the Hilbert space $L_2(\mathbb{R}^N)$. This Hilbert space, in common with the Hilbert space for the single quantum oscillator, is infinite dimensional. However, it can be truncated a natural way by restricting attention to values of $n \in \mathbb{Z}^N$, and using $\log_2 N$ qubits to label the values of n. These states not only form the basis for the Hilbert space $L_2(\mathbb{C}^N)$, but also form a space of holomorphic functions that live on a complex torus $\mathbb{C}^N/\Lambda$, where Λ is a 2D lattice and $\mu, \nu \in \mathbb{Z}^N/(\mathbb{Z}^N/m)$ where $N = m^2$. This Hilbert space, in common with the Hilbert space for the single quantum oscillator, is infinite dimensional. However, it can be truncated a natural way [by first projecting the sum (6.15) onto the "radial wave functions" for an oscillator array to form a representation for any square integrable function of $r = |z|$ in $\mathbb{C}^N$:

$$f(r) = \sum_{k=0}^{\infty} \left(\int_0^{\infty} f(s)\varphi_k(s)s^{2n+1}ds \right) \varphi_k(r), \qquad (6.18)$$

where the $\varphi_k(r) = [\frac{2^{-n}k!}{(k+n)!}]^{\frac{1}{2}} r^2 e^{-\frac{r^2}{4}} L_k^n(r)$ are the generalized Laguerre functions which also appear in elementary quantum mechanics [14]. The expansion (6.18) can also be expressed as a projection:

$$P_k f(z) = \sum_{|\beta|=k} \sum_{\alpha} (f, \Phi_{\alpha,\beta})\Phi_{\alpha,\beta},$$

where P_k projects f onto the space spanned by $\{\Phi_{\alpha,\beta}, \alpha, \beta \in \mathbb{Z}^N, |\beta| = k\}$. One interesting thing about this expansion is that it can be naturally truncated to a finite sum

$$Q_k f(z) = \sum_{|\mu|=k} (f, \Phi_{\mu-m,\mu})\Phi_{\mu-m,\mu}, \qquad (6.19)$$

which transforms homogeneously under the torus subgroup of the group of $n \times n$ unitary matrices acting on an n-dimensional Hilbert space; i.e. under $z \to e^{i\theta}z$

$$f(e^{i\theta}z) = \sum_m Q_k f(z)e^{im \cdot \theta}. \qquad (6.20)$$

The finite dimensional Hilbert space spanned by $\{\Phi_{\alpha,\beta}, \alpha\beta \in \mathbb{Z}^N, |\beta| = k\}$ is evidently a close relative of the Θ-functions that were used [54–57] to represent Riemann surfaces and which play a central role in finding exact solutions.

The Θ-functions of interest in connection with Riemann surfaces are defined by the relation [54–56]:

$$\theta \begin{bmatrix} a \\ b \end{bmatrix} (z) = e^{2\pi i a \cdot b} X(b) Z(a) \theta(z), \tag{6.21}$$

where a, b are real vectors, and X and Z are the Weyl–Heisenberg group elements (see Appendix C)

$$X(y) = \exp(-iy \cdot \partial/\partial \xi) \text{ and } Z(x) = \exp(ix \cdot \xi). \tag{6.22}$$

In the case where the shifts x and y are restricted to the integers, mod n and z lie on a complex torus $\mathbb{C}^g/\Lambda$, where Λ is a 2g dimensional lattice. The indexed Θ-functions (6.18) were originally introduced [54] by Solomon Lefshetz as the coordinates of a Riemann surface embedded in flat projective space P^N. (This embedding is of particular importance in mathematics because it means that Riemann surfaces are "algebraic varieties".) Remarkably, the set of functions defined in Eq. (6.21) form a "reproducing" kernel space (cf. [99]) of dimension n^{2g}. The term reproducing means that they are the eigenfunctions for the defining kernel, which in the case of (6.21) is the closed string theory propagator used in relativistic string theory [100]. These functions can also can be constructed [55] by first identifying their value at a reference point $z = 0$, and then using the Weyl–Heisenberg shift operators, Eq. (6.2) to define their values over the entire Riemann surface. It was Lefshetz's discovery of the embedding of Riemann surfaces in projective space using these functions that allowed quantum mechanics to emerge from algebraic geometry. (For a detailed discussion of theta functions with characteristics, see Griffiths and Harris's *Principles of Algebraic Geometry* [54] or Mumford's more succinct *Tata on Theta* [56].)

6.2. Helstrom's Theorem and Universal Hilbert Spaces

The primary task of quantum pattern recognition might be viewed [99] as choosing a feature Hilbert space $\mathcal{H}_F$ and a map $\mathbf{x} \to \Psi(\mathbf{x})$ from input data to the space $\mathcal{H}_F$ such that the features represented in the data are easily distinguished. Because the number of quantum states that can be represented with even a finite set of basis states is literally infinite, it might seem that there would be an enormous advantage to storing data features as quantum states. However, this is probably a chimera because one must take into account that there are strict limits in how much information can be stored in quantum states. The key to understanding this is Helstrom's theorem [102], which places strict limits on the distinguishability of two quantum states. Helstrom's theorem plays a role in quantum Bayesian inference that is analogous to the singular role that the Neyman–Pearson test plays in classical Bayesian approaches to data interpretation. One of the advantages of quantum information processing is that as a consequence of Helstrom's theorem, one is able to immediately attach information theoretic significance to the data features regardless of whether these features are Gaussian distributed variables.

One of the enigmas of quantum information processing is whether it is possible how to encode experimental data as quantum states. For example, if one wants to know how many measurements are needed to distinguish two Gaussian distributed variables, one only needs to know the estimated mean and variance for the two variables in order to determine for example the probability of false alarm (PFA) were really the same even when the measurement suggested they were different. However, in quantum mechanics the wave functions themselves are a deterministic rather than probabilistic quantity. Therefore, in quantum mechanics there is no automatic way of associating information with a state in Hilbert space. Nevertheless, there is a simple and universal way for estimating the PFA for quantum measurements. Namely, the probability of "false alarms" is elegantly provided by Helstrom's theorem:

$$\text{PFA} = 1 - \sqrt{1 - \eta}, \qquad (6.23)$$

where $\eta = [< \Psi_1 | \Psi_2 >]^2$. This estimate for the PFA is independent of the number or type of measurements. Thus, Helstrom's theorem does provide a limitation on how well quantum measurements can reproduce Bayes's conditional probabilities. However, in practice the statistical uncertainties associated with weak measurements typically obscure this limitation. On the other hand, as noted in the introduction, for the most part we are going to restrict our attention to weak measurements, which allows Bayesian conditional probabilities to appear in a completely natural way.

The underlying presence of quantum mechanics in any real example of observations is revealed by the fundamental limitation (6.23) on any measurement. In general, machine learning recognizes patterns [4,7] by constructing a nonlinear mapping, $x \to z(x)$, between a set of input data vectors $\{x^{(n)}\}$ and a smooth interpolation function $z(x)$ in feature space $\mathcal{H}_Z$. As noted in Ref. [6], in many cases the interpolation function $z(x)$ can often be constructed as the least square estimator based on a correlation function for the data. In conventional machine learning, e.g. using neural networks, the central task can be viewed as the hierarchical construction of the nonlinear map $x \to z(x)$ by using the kernel matrix $K(x,y)$ which represents the features at one level to construct the kernel matrix at the next level. At each stage of machine learning, it is usually assumed that the kernel describing correlations in feature space is "reproducing" in the sense that its eigenvectors are the basis for feature space $\mathcal{H}_F$.

Following the legacy of Wiener's 1958 essay [97], interest has recently increased in using quantum eigenstates and associated kernels to represent the feature spaces (see e.g. Schuld *et al.* [98,99]). For example, the spatial part of the energy eigenvalue states for a quantum harmonic oscillator has the form

$$\phi_k(x) = \exp(-(c-a)x^2)H_k(\sqrt{2c}x), \tag{6.24}$$

where $a = 1/4\sigma^2$ and $c = (a^2 + 1/4l^2\sigma^2)^{1/2}$, and $H_k(x)$ is a Hermite polynomial. As it happens, the analytic properties of these eigenstates make them natural candidates to stand-in for GPs. The reproducing kernel in this case is

$$K(xx') = \exp(-(x-x')^2/2l^2). \tag{6.25}$$

As it happens, the usefulness (6.24) for representing the features in datasets of practical importance has been recognized for some time. As discussed in Chapter 3, this kernel function provides a prediction model for feature labels $l^{tn)}$ attached to a dataset $\{x^{(n)}\}$, when the least square regression model $z(x)$ for these labels is modeled as a sum of "radial basis" Gaussian functions centered at discrete points. The eigenfunctions of the kernel $\exp(-|x-x'|^2/2\sigma^2)$, and when defined in this manner the basis for the feature Hilbert space, $\mathcal{H}_F$ turns out to have the form of a Gaussian exponential times Hermite polynomials.

The reproducing kernel for two quantum harmonic oscillators can be written in the form

$$K(x, y) = \sum_{n_1 n_2} \varphi_j(x)\varphi_j(y), \qquad (6.26)$$

where ϕ_j is an eigenfunction of the Hill–Schrodinger operator $-\frac{\partial^2}{\partial x^2} - \frac{\partial^2}{\partial y^2} + \frac{1}{2}(x^2 + y^2)$ function. (This operator first appeared in the 19th century in connection with Hill's theory of the stability of Lagrange triangles, but reappeared in 1925 in connection with Schrodinger's equation for an "upside down" 2D quantum oscillator). It turns out that the radial part of wave function for a 2D quantum oscillator involves a generalized Laguerre polynomial L_M^n that is closely related to generalized Laguerre polynomial that appears in the radial wave functions for the 2D hydrogen atom problem [14]. This brings us full circle back to the problem that originally inspired Bayes and Gauss; i.e. finding the orbital parameters for astronomical objects moving under the influence of a $1/r$ potential. It is worth keeping this in mind because this suggests that the model selection problem that attracted Gauss's interest, namely assigning multiple solar system objects to distinctive orbits, might also be treated as a quantum problem for multiple oscillators.

This focus on the 2D harmonic oscillator permits us an easy segue to another very important Hilbert space related to the quantum theory of angular momentum. Following the epochal 1922 discovery by Stern and Gerlach of "spatial quantization" [131], Wigner and Racah [113] developed a beautiful formalism for describing angular momentum states in quantum mechanics. These states are of interest for quantum machine learning because of a connection between the energy eigenstates (Fock states) of a quantum oscillator and quantum angular momentum states discovered by Julian Schwinger (when he was a graduate student!). In *On Angular Momentum*, Schwinger describes a very elegant way of constructing the quantum angular momentum operators $\hat{J}_x$, $\hat{J}_y$, and $\hat{J}_z$ as well as the Wigner–Racah algebra [113] for representing vector sums of angular momentum in terms of the annihilation and creation operators for a 2D quantum harmonic oscillator. (These notes are unpublished, but a brief summary can be found in [112].) Schwinger's construction of these states is based on representing the quantum angular momentum operators in terms of the raising and lowering operators for the number states of a 2D quantum harmonic oscillator:

$$J \equiv \hbar \sum_{\varsigma,\varsigma'=\pm} a_\varsigma^+ \left\langle \varsigma \left| \frac{\sigma}{2} \right| \varsigma' \right\rangle a_{\varsigma'}$$

$$[a_\varsigma, a_{\varsigma'}] = [a_\varsigma^+, a_{\varsigma'}^+] = 0 \quad \left[a_{\varsigma'}, a_\xi^+\right] = \delta_{\varsigma\varsigma'} \qquad (6.27)$$

$$J_+ = \frac{\hbar}{2} a_1^+ a_2, \quad J_- = \frac{\hbar}{2} a_2^+ a_1, \quad J_3 = \frac{\hbar}{2}(a_1^+ a_1 - a_2^+ a_2).$$

One potential advantage of using the Fock states for a 2D oscillator to represent quantum angular momentum states is that superconducting quantum oscillators provide an analog method for representing these states.

6.3. Measurement-based Quantum Computation

Although our presentation has for the most part ignored the extensive literature on qubit quantum computing, there is one development in qubit quantum computing that mirrors our approach to Bayesian inference: the "measurement-based quantum computing" formalism of Raussendorf and Briegel [132]. Our approach to finding the optimum strategies for Bayesian search and model selection problems by encoding both observational data and the conditional probabilities used in Bayesian inference as self-organized quantum states is very similar in spirit to using measurements of entangled states of qubits to carry out quantum computations. In Ref. [131], it was shown that essentially all quantum computations that have been contemplated using qubit quantum circuits can also be carried out by making measurements of qubit states in a 2D array whose quantum states have become entangled by applying a controlled phase gate $C_Z = \exp\left(-i\frac{\pi}{4}\sum_{<i,j>} \sigma_i^z \sigma_j^z\right)$ between qubits on neighbor nodes. Such controlled phase gates can be realized naturally by allowing an Ising spin-like interaction between neighboring qubits to act for time intervals analogous to the Rabi time for spin flip in a magnetic field. As a simple illustration of how measurement-based quantum computing works for qubits, we consider the problem of teleporting a bipartite qubit state of the form $(\alpha_1|0> + \beta_1|1>)(\alpha_2|0> + \beta_2|1>)$ from one location to another.

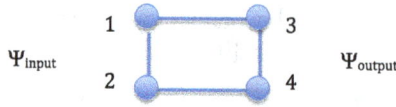

The teleportation is accomplished using the controlled phase gate

$$C_Z = \exp i \left(\frac{\pi}{2}\sigma_z \otimes \frac{\pi}{2}\sigma_z \right), \qquad (6.28)$$

where $\sigma_z |i> = (-1)^i |i>$ to entangle the states 1 and 3 and 2 and 4. After initializing the qubits in states 1 and 2 with arbitrary initial states and the states in 3 and 4 with the Hadarmard states $|0> +|1>$, followed by the application of C_Z, the wave function has the form

$$(\alpha_1 |0>_1 \sigma^z_{(3)} + \beta_1 |1>_1)(|0>_3 +|1>_3)$$
$$\otimes (\alpha_2 |0>_2 \sigma^z_{(4)} + \beta_2 |1>_2)(|0>_4 +|1>_4). \qquad (6.29)$$

After transforming to the conjugate basis $|\pm> = (|0> \pm |1>)/\sqrt{2}$, this wave function has the form

$$\sum_{s_i=0,1} \exp\left(-i\frac{\pi}{2}s_i\sigma^x_3\right)(\alpha_1 |->_3 +\beta_1 |+>_3)$$

$$\otimes \sum_{s_i=0,1} \exp\left(-i\frac{\pi}{2}s_i\sigma^x_4\right)(\alpha_1 |->_4 +\beta_2 |+>_4). \qquad (6.30)$$

Then measuring the eigenvalues of σ^x_1 and σ^x_2 yields the initial wave function defined on the 1 and 2 nodes teleported to the 3 and 4 nodes.

As an illustration of the potential usefulness of this type of scheme for Bayesian searches, we consider the "Monty Hall" search problem where the location of an object of interest within a linear array of boxes is being sought [5]. The 2D quantum oscillator array we envision using for this problem is illustrated in Fig. 6.1.

In this cartoon, each node of the middle layer consists of either a single quantum oscillator plus a qubit or a pair of quantum oscillators. This layer is the "quantum computer" which we use to find the location of the hidden object. The quantum computations can be carried out with either N levels of the single quantum oscillator or with $[N/2] + 1$ levels in each oscillator of a pair of quantum

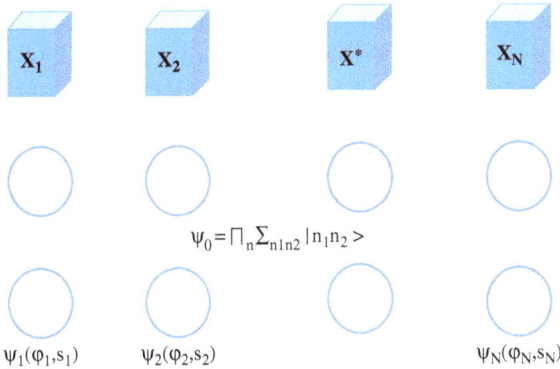

Fig. 6.1.   Quantum scheme for solving the Monty Hall problem.

oscillators. The bottom layer contains the wave-functions representing the observational data which consists of a sequence of choices for locations to be searched in boxes and the results of the examination of the boxes at each location, while the upper "hidden" layer contains information about possible explanations for the input data. One might envision a quantum version of the Monty Hall problem consisting of pairs of quantum oscillator states to encode both the locations to be searched and the posterior probabilities for finding the object in each location. The computation consists of introducing prior information about possible models for the input data into the hidden layer by teleportation, and then using a sequence of observations to relax the hidden layer wave function to find at each step of the search the models which best match the data gathered up to that time. Whereas in the case of classical Bayesian searches it is typically an ad hoc assumption that the location of the object being sought will be identified with a high probability in N or fewer steps, in a quantum mechanical approach this result is a natural consequence of the formalism. Indeed, to identify the location of an object that classically might be in an infinite number of locations, in quantum mechanics approaches success in N steps appears as a natural result of "space quantization". As noted in the last section the position variable operator Q_i associated with a 2D oscillator has a natural interpretation as the z-component of the angular momentum operator $\boldsymbol{J}$. This opens the door to using spherical harmonic functions to define quantized locations on a sphere. Although any location on the surface of a sphere can be precisely defined using an infinite sum of spherical harmonic functions, one must keep in mind that practical quantum computations are only possible with a finite number of basis states.

If the presence or absence of the object at a particular location cannot be exactly determined, then the spin part of the wave function in Eq. (6.31) will have to be replaced with a superposition $\alpha|1> +\beta|0>$ of the $s_i = 1$, and $s_i = 0$ states which represents the measurement uncertainty. The wave function describing the input data accumulated up to the time of the nth step of the search will therefore have the form

$$\Psi_{in} = \prod_{i=1}^{i=n} \sum_k \lambda_i^k |\psi(\phi - \phi_k), \alpha_k|0> + \beta_k|1 \gg \qquad (6.31)$$

for $i = 1, \ldots, n$ where the $\lambda_i^n = 0$ or 1 is a parameter which describes which of the N possible locations is observed at the ith step of the search. The product extends over all observations up to the time of the measurement. Adopting the same convention used in measurement-based quantum computing we will assume that the initial wave functions for the hidden layer nodes are the zero modes $|\psi(\phi), | + >)$; i.e.

$$\Psi_0 = \left| \sum_k \left| \psi(\phi - \phi_k), + > \right) >_1 \prod_{i=2} \right| (\psi(\phi), + >) >_i,$$

where $\psi(\phi)$ is the best approximation to a delta function that can be achieved with spherical harmonic functions with l limited to l_{max} and $|+ >= |1 >_I + |0 >_I$. Our Bayesian search will be implemented by first delocalizing within the hidden layer these models using the controlled phase gate of the form (6.30) to entangle the nodal zero mode wave functions in nodes $i = 1,2,3,4$ (corresponding to $N = 4$) of the hidden layer. This leads to an entangled wave function for the hidden layer:

$$\Psi_H = \frac{1}{2} \left(\left| \sum_k |\psi(\phi - \phi_k) e^{iL_1^z L_2^z}, \middle| 0 > \sigma_2^z + \middle| 1 > \right) \right.$$

$$\times \prod_{h=2}^{h=N} \psi(\phi)(|0 > \sigma_{h+1}^z + |1 >)_h, \qquad (6.32)$$

where $\sigma_{N+1}^z = 1$. Expanding the r.h.s of (6.32) after entanglement, the search proceeds by measuring at each step of the search the nodal wave functions in a basis corresponding to the input data. In particular, we propose to measure the hidden layer nodes in a basis $\psi(\phi - \phi_i)B_i(\theta_i)$, (we assume here that the measurement at each node can readily determine which "box" is being observed), where

$$B_i(\alpha_i) = \left(\frac{\cos\theta_i/2|1> -\sin\theta_i/2|0>}{\sqrt{2}}, \frac{\sin\theta_i/2|1> +\cos\theta_i/2|0>}{\sqrt{2}} \right)$$

is a two-component spin state that represents the measured spinor state $\alpha|1>_i +\beta|0>_i$, where θ is the usual Cayley–Klein parameter for a spinor state corresponding to a rotation about the y-axis of the initial state $|1> + |0>$. After the measurement of a node state the result is either a spin variable s_1 equal to the first component of the two-component spin state $B_i(\theta_i)$, or a spin variable s_2 equal to the second component of the state $B(\theta_i)$. After measuring the nodal wave functions in the wave function (6.32), the hidden layer has the form

$$= \frac{1}{2}(|\psi_1 s_1(\theta_1)>_1|\psi_0 s_2>_2|\psi_0 s_3>_3 X(s_1, s_2, s_3)$$

$$\times \sum_k e^{i(\theta_1 \sigma_k^z/2 + \phi_1 L_k^z)}|\psi_0>_4),$$

where X is a known correction factor depending on the measured values of s_1, s_2, and s_3 while the unitary operator $U = e^{(\theta_k \sigma_k^z/2 + \phi_k L_k^z)}$ appearing in the node 4 brings all the model wave functions closer to the input wave function at step 1. Introducing additional input data pushes the model states toward the input states in the order they appear in the search strategy. Our quantum search amounts to evolving the initial wave function (6.31) in such a way that as measurement data are accumulated, the wave function in node 4 of the hidden layer approaches a wave function representing the locations searched and the posterior probabilities for finding the object at various locations at each step of the search. Although this excursion into qubit information processing does not provide much of an improvement over classical search algorithms, it does illustrate the importance of using measurements of a "momentum" variable to guide the evolution of the wave function.

Chapter 7

Quantum Self-organization

7.1. Pontryagin Control and Quantum Criticality

As has been emphasized by Kappen [31], if the innovation noise in a controlled system is below a certain critical level, then the Bellman control theory reverts to a deterministic form known as Pontryagin control [86,87]. That is, in the limit when the Bellman function is optimized and the stochastic control costs are limited, Pontryagin control takes over. In this limit, solutions of optimal control problems are typically represented as smooth flows in $(x, \dot{x})$ space, where system and control variables are continuously well defined in the same way as the position and momenta variables in classical mechanics. The phase space diagram of the moon lander problem discussed in Chapter 4 is a nice illustration of the qualitative nature of the solution. In principle, these streamlines can be found by solving the Hamilton–Jacobi–Bellman (HJB) equation [18,20] for various initial values x_0. The HJB equation is similar to the classical Hamilton–Jacobi equation (see e.g. [88]), except that the velocity $\dot{x}$ is identified as a control variable u. A generic feature of Pontryagin control is that as a function of initial conditions the ensemble of controlled paths parameterized by position and control variables resemble flows of a 2D fluid. In a quantum regime, such flows can in turn be represented as solutions of the Gross–Pitaevski equation [133] describing the collective state of a multi-boson system. A potential advantage then for using quantum wave function to represent a classical stochastic game (x, u) space is replaced by a single coherent state.

Because of its determinism, classical mechanics does not offer a mechanism for storing or processing information. In contrast, quantum dynamics does offer such a possibility because of the direct connection between the Schrodinger equation and information theory that was originally discovered by David Bohm [134] as a result of his attempt to reformulate quantum mechanics as a statistical theory of hidden variables. Despite this ill-fated expedition, the equations he wrote down do resonate with the goal of this book. Bohm reformulated the Schrodinger equation as a nonlinear equation for the quantum phase that looks suspiciously like the HJB equation that plays a central role in control theory (see e.g. [18,20]). Our description of Bohm's result follows Frieden [135] and Reginatto [136]. The Schrodinger equation in a "hyperbolic" space-time; i.e. a space-time that is layered in such a way as to admit the universal time, has the form

$$ i\hbar \frac{\partial \psi}{\partial t} = -\frac{\hbar^2}{2} \sum_{i,j}^{d} g^{ij} \frac{\partial^2 \psi}{\partial x^i \partial x^j} + v(\mathbf{x}), \tag{7.1} $$

where the x variables represent the positions of particles and $v(\mathbf{x})$ is the potential. They then assume that the wave function $\psi(x)$ has the form

$$ \psi(x) = P^{1/2} e^{iS/\hbar}, \tag{7.2} $$

where P is a real function chosen so that integral of $|\psi(x)|^2$ over all space equals 1. Separating out the real and imaginary parts of Eq. (7.13) leads to two coupled nonlinear equations for P and S [135–136]:

$$ \frac{\partial P}{\partial t} = -\sum_{i,j=1}^{d} g^{ij} \frac{\partial}{\partial x^i} \left(P \frac{\partial S}{\partial x^j} \right) \tag{7.3} $$

$$ \frac{\partial S}{\partial t} = -\frac{1}{2} \sum_{i,j}^{d} \left(g^{ij} \frac{\partial S}{\partial x^i} \frac{\partial S}{\partial x^j} + v(\mathbf{x}) \right. $$

$$ \left. -\frac{\hbar^2}{8} \left(\frac{2}{P} \frac{\partial^2 P}{\partial x^i \partial x^j} - \frac{1}{P^2} \frac{\partial P}{\partial x^i} \frac{\partial P}{\partial x^j} \right) \right). \tag{7.4} $$

Rewriting Schrodinger's equation as in Eqs. (7.3) and (7.4) has turned out to be of interest in a number of contexts; e.g. its similarity to both the NLS equation and the HBJ equation, which suggests a connection with stochastic control theory (cf. [67]). The nondeterministic nature of quantum dynamics is signaled by the appearance of the 3rd term on the r.h.s of Eq. (7.4) — which is known as the "Bohm potential". In Refs. [133,134], it is noted that the Bohm potential is an example of "Fisher information", which is closely related to the Kullback–Leibler (KL) divergence [27] that is an information theoretic measure of how well an estimated probability density matches a "true" probability density, e.g. a conditional probability that one might obtain from Bayes's formula. When an estimate of a desired probability density is improved, for example as a result of an observation, then the KL divergence changes by an amount equal to the information that has been gained as a result of the measurement. When the Bellman function is optimized, no more information gathering is possible, and the dynamics is described by Pontryagin control.

When the number of particles is a large number and the wave function $\Psi(x_1, \ldots, x_N)$ is required to be symmetric under interchange of the coordinates of any two bosons, then the Schrodinger equation (7.11) can often be replaced by the Gross–Pitaevskii description [133] of a cloud of bosons with off-diagonal long range order (ODLRO). The Gross–Pitaevskii equation for the order parameter can be derived as the Euler–Lagrange equation for a classical Lagrangian:

$$\mathfrak{L} = \psi^* \left(i\hbar \frac{\partial}{\partial t} + \mu \right) \psi - \frac{\hbar^2}{2m} |\nabla \psi|^2 - U|\psi|^2, \tag{7.5}$$

where μ is the chemical potential. Near a critical point where the speed of sound in the boson fluid vanishes (which implies the Pontryagin condition), the effective Lagrangian is

$$\mathfrak{L}_{crit} = \psi^* \left(i\hbar \frac{\partial}{\partial t} + \mu \right) \psi - \frac{\hbar^2}{2m} (|\psi| - \psi_0)^4. \tag{7.6}$$

What we would like to emphasize here is that at a critical point where the speed of sound in the quantum fluid vanishes "time stands still", which is exactly the condition that the control Hamiltonian $H_c = 0$, which corresponds to Bellman optimization devolving to the Pontryagin "maximum principle". The point of departure for our proposed quantum approach to Pontryagin control is to replace the Schrodinger wave function that appears in Eq. (7.11) with a wave function satisfying the Gross–Pitaevskii equation for the order parameter for a bosonic fluid:

$$i\hbar \frac{\partial \psi}{\partial t} = -\frac{\hbar^2}{2} m \nabla^2 \psi + [U(|\psi|^2) + \phi(x)]\psi, \tag{7.7}$$

where $\psi(x, t)$ is the complex valued order parameter whose absolute square is the spatial density of bosons in the cloud, and $\phi(x)$ is an external potential (e.g. due to an external spatially varying gravitational field acting on the bosonic cloud in an atomic "fountain"). The particle density and velocity field corresponding to (7.5) are smooth and satisfy the Euler–Khalatnikov equation [137]:

$$m \frac{\partial v(x)}{\partial t} = -\nabla \left\{ \frac{1}{2} m v^2 - \frac{\hbar^2}{4m} \left[\nabla^2 \ln\rho + \frac{1}{2}(\nabla \ln\rho)^2 \right] \right\}. \tag{7.8}$$

The semi-classical fluid dynamics described by the Eq. (7.6) have the characteristic property that the flows described by these equations resemble Eulerian flow around obstacles. (As a historical aside, it took observations of objects dropped from the Eiffel Tower to realize that ordinary hydrodynamics was not Eulerian.) What makes this work as a model for machine learning is that the phase of the order parameter $\psi(x, t)$ along the streamlines of the flow (7.6) satisfies the HJB equation, where the potential $U(x)$ plays the role of the reward function.

The bosonic fluid described by Eq. (7.7) can also be regarded as a multi-boson system defined by a Hamiltonian of the form

$$H_c = -\frac{\hbar^2}{2m} \sum_{i=1}^{N} \partial_i^2 + U(x_1, \ldots, x_N), \tag{7.9}$$

An interesting sidelight for the Gross–Pitaevskii model for control is that the theory corresponding to Eq. (7.9) can also be interpreted [138] as a quantum theory of space-time near a black hole horizon. There is no consensus at the present time as to what a quantum theory of gravity looks like; however, the Gross–Pitaevskii model, defined by Eq. (7.9), may cut the "Gordian Knot" as to the quantum theory meaning of an event horizon including a theory of black hole entropy [139]. Of course, this begs the question as to why some interesting RL problem might be mapped onto such a model, but our Gross–Pitaevskii model suggests that one way to understand optimization of the Bellman function is to observe the time history of the order parameter falling in a gravitational field as a function of altitude:

$$i\hbar\frac{\partial\psi}{\partial t} = -\frac{\hbar^2}{2m}\nabla^2\psi + [U'(|\psi|^2) - g(t)h(x)]\psi, \qquad (7.10)$$

If the altitudes h_i of atoms in the cloud and their velocities v_i are controlled by varying the acceleration of gravity g, then the optimal point is the altitude h where the speed of sound vanishes, and the phase of the order parameter has a stationary point.

The moon lander problem described in Chapter 4 illustrates what the solution to Eq. (7.10) might look like. In this case, turning on the rocket thrust of the moon lander can be emulated in a cloud of bosons falling in a gravitational field by demanding that on reaching a surface the bosons should come to rest. Another advantage of visualizing the system being controlled as a quantum fluid is that fluid control is susceptible to H_∞ control [119], which provides an entre [84] for quantum self-organization to the theory of games.

7.2. Quantum Theory of Innovations

An obvious challenge for the quantum version of the Durbin–Willshaw set-up [1] for solving the traveling salesman problem (TSP) is how to emulate the nonlinear couplings between the locations where the salesman stops and the observed locations of the cities. An attendant problem is how to model using quantum mechanics the evolution of the receptive fields for the cities in Fig. 2.1; i.e. the set of cities connected by dashed lines to each city. If these receptive fields overlap, then as observed the salesman's itinerary regresses toward the true trajectory of the topology of the connections between the station-stops and the cities may change to optimize the length of the salesman's path (or Bellman value function in the case of feedback control or RL). One of the pleasant consequences of using quantum paths in a path integral to represent the model paths in the TSP is that the receptive fields and nonlinear springs associated with the

$d(i, \mu)$ in the Durbin–Willshaw set-up are replaced by a continuous local force acting on a particle salesman traveling along a model path which only depends on the true velocity of the salesman at any given time. The existence of an instantaneous influence (that may however depend on the previous history of a system) acting to modify the current state of system based on observations of signals from the system is a universal feature of Bayesian feedback control and reinforcement learning.

The quantum solution of the TSP introduced in Chapter 2 is a good illustration of how this process works, and in addition how the quantum theory of propagation of a particle in a magnetic field can be used to emulate how the differences between the estimated state of a system and its actual state can be reduced by the introduction of a control variable. As it turns out, there is a perfect correspondence between the deviations in the velocity of particle along a trial path from the classical velocity and the action in the Feynman quantum amplitude for motion along the path, which has the form

$$K(t) = \int \exp\left[\frac{i}{\hbar} \left(S_{cl}[x(t')] + \Delta S \lfloor [x(t')] \rfloor\right)\right] Dx(t'). \qquad (7.11)$$

The factor containing $S_{cl} = (x_b - x_a)^2/2\left((t_b - t_a)\right)$ describes the classical motion of a free particle between cities (see [47]), while the factor containing ΔS describes the deviation of the observed motion of the "quantum salesman" from the classical path:

$$F(t - t_0) = \int \exp\left(\frac{i}{\hbar} \int_{t_0}^{t} \left(\frac{m}{2}|\dot{x} - v(t')|^2\right) dt\right) Dy(t'), \qquad (7.12)$$

where the classical velocity $v(x)$ is defined for all x by the actual motion of the salesman, and $y(t) = x(t) - x_{cl}(t)$ is the deviation of the Feynman path from the classical path. It is a crucial observation for us that the action functional in (7.12) can also be expressed as the effective action function for the motion of a classical particle in a magnetic field [138].

$$\Delta S \lfloor [x(t)] \rfloor = \int_{t_0}^{t} \left(\frac{m}{2}\dot{x}^2 + \dot{x}A(x)\right) dt, \qquad (7.13)$$

where $\vec{A}(x) \equiv m\vec{v}(x)$ is formally the familiar vector potential for a magnetic field $B = curl A$. In the context of Eq. (7.13), it is a "control variable" is just the velocity of the salesman at that particular time. Thus, we arrive at the surprising result that the innovation *GP* is equivalent to the quantum motion of a particle in a position-dependent magnetic field.

As a hint we are on the right track to finding a quantum translation for optimal control, the quantity inside the parenthesis in (7.13) has the same form (apart form a sign) as the control Hamiltonian for Pontryagin control discussed in Chapter 4, with the proviso that the control variable $A(x)$ is now interpretable as a vector potential. Applying Eq. (7.13) to solve the traveling salesman problem seems straightforward; namely one considers the classical limit, where because of cancellation of phases the path integral becomes focused on the shortest path length.

Restricting attention to quantum deviations within a region surrounding the straight-line classical path between cities with locations ζ_b and ζ_a with area $|\zeta_b - \zeta_a|\Delta y$ then according to Landau (see *Quantum Mechanics* [15]) the number of quantum states needed to describe the deviation is $B|\zeta_b - \zeta_a|\Delta y/2\pi\hbar$, each of which has the form $\exp\{-2\pi\hbar(z - z_i)^2/B\}$; i.e. a localized Gaussian density. Evidently, (omitting a normalization factor) the ground state describing the localization of the salesman within the region corresponding to the deviations from the classical path has the form

$$\Psi = \exp\{-2\pi\hbar S/B\}, \tag{7.14}$$

where S is the area of the region between the salesman's path connecting the square dots in Fig. 2.1 and the observations of the salesman's path linking the round dots representing the cities. $B = \text{curl}A$ is assumed to be constant in this region. Of course, this formula only makes sense on a Riemann surface because in the plane the salesman's itinerary will in general be a self-intersecting graph. Equation (7.14) is completely consistent with our conjecture [36] that the area S plays the same role as the Bellman function in optimal control, i.e. the area S represents the information regarding the salesman's itinerary that has been lost as a result of environmental noise. In other words, in our quantum version of the TSP the innovation noise is just a consequence of using a path integral with the Nambu-like action for a string [100] to describe the observations. When the deviations of the quantum path from the classical path are limited such that the path is not self-intersecting, then Bellman optimization is equivalent to minimizing the area in Eq. (7.14). This is a very pleasant result in the sense that the Bellman function, which represents a loss of information, can now be interpreted as the area spanned by the innovation. Of course, when the lost information about the salesman's movements due to the presence of the control variable $\vec{A}(x)$ is recovered, the salesman becomes localized on the classical path.

Although the action in Eq. (7.12) had its origin in our approach to the TSP, the appearance of the vector potential $\vec{A}(x)$ suggests that a physical analog realization of innovation for feedback control would require magnetic fields. Serendipitously, it was discovered by Duncan Haldane [140] that an effective magnetic field can spontaneously appear on the surface of certain atomic lattices with strong spin orbit interactions. Somewhat later a class of materials, topological insulators, was discovered with exactly this property [140]. Here we just note the possibility that the quantum innovation, Eq. (7.12), might have an analog realization involving topological insulators or superconductors. Spin orbit effects in topological insulators are controlled by a parameter $1/\kappa =$ spin orbit influence length/atomic spacing and in TIs can be represented as the Chern–Simons out of plane "magnetic field:

$$B_{CS} = -\frac{e}{\kappa}\psi^*\psi$$

and an in-plane electric field

$$E_{CS}^i = \frac{e}{\hbar v_F}\varepsilon^{ij}j^i.$$

The 2D Schrodinger equation describing the motion of a 2D quantum fluid of particles interacting via both a point-like interaction and the gauge potentials for these Chern–Simons fields has the following form [142]:

$$i\hbar\frac{\partial\psi}{\partial t} = -\frac{1}{2m}D^2\psi + eA_0\psi - g|\psi|^2\psi, \tag{7.15}$$

where $D_\alpha = \partial_\alpha - i(e/\hbar c)A_\alpha$ and m is an inertia parameter. The gauge fields A_0 and A_α do not satisfy Maxwell's equations, but instead are determined self-consistently from the equations for Chern–Simons electrodynamics in $2+1$ dimensions. In the presence of a uniform 2D electric field E, the current has the same form as the Hall current for a magnetic field perpendicular to the plane:

$$j_{\alpha\beta} = \sigma_H \epsilon_{\alpha\beta\gamma} E_\gamma,$$

where σ_H is the "Hall conductivity". Neglecting spatial variations in the electric field, the usual Gauss's law will be replaced by the Chern–Simons equation

$$B = -\frac{e}{\kappa}\varrho,$$

where B is the strength of an effective magnetic field whose direction is perpendicular to the surface, $\varrho = \psi^*\psi$, and $1/|\kappa|$ is an inverse length with $\sigma_H = v_F\kappa/e$, and v_F is the Fermi velocity on the surface of the TI. The control parameter appears as the vector potential that appears in the covariant derivatives D_x and D_y that appear in Eq. (7.15). Physically, the appearance of this vector potential is associated with the effect of spin orbit coupling between spin polarized charge carriers moving on the surface of the TI and the nuclei of high Z atoms just below the surface.

Although the magnitude of the surface vector potential in the surface of a topological insulator cannot be controlled, the charge carrier flows implied by Eq. (7.14) are collectively very smooth. The Hamiltonian corresponding to Eq. (7.15) is

$$H = \int d^2x \left\{ \frac{\hbar^2}{2m} |(D_x \pm iD_y)\psi|^2 - \frac{1}{2}\left(g - \frac{e^2\hbar}{mv_F\,|\kappa|}\right)\varrho^2 \right\}.$$
$$(7.16)$$

As in optimal control theory, we are interested in the Pontryagin limit $H = 0$. Simple analytic expressions for these zero modes can be found if we assume

$$g = \pm e^2\hbar/mc\kappa,$$

the equation for zero modes $H = 0$ reduces to

$$(D_x \pm iD_y)\psi = 0. \qquad (7.17)$$

This equation has the simple analytical solution

$$\psi(x) = e^{ikx}, \quad \nabla\phi = k\hat{x}, \quad \vec{j} = -\frac{e}{mv_F}\vec{A} = \rho\frac{\hbar k}{m}\hat{x}. \qquad (7.18)$$

We see here that for small deviations between the model path and the salesman's path the holomorphic flows satisfying (7.17) closely mimic the TSP innovation. This is an illustration of the uniform convergence appearance of phase space parameters in Pontryagin control. Equation (7.17) can also be written as a Liouville equation:

$$\nabla^2 \ln \rho = \pm\frac{2e^2}{\hbar v_F \kappa}, \qquad (7.19)$$

which has vortex-like solutions [142]. The bottom line is that current flows on the surface of a topological superconductor [141] may be a way of representing Kailath's innovations in an analog quantum device [143].

7.3.   Quantum Helmholtz Machine

The aim of the wake–sleep algorithm [9] for training the Helmholtz machine is to produce joint probability distributions for the ensemble of Ising spins (states = $\{+1, -1\}$) which minimizes the information costs of representing a set of observations. One of the key ideas behind the Helmholtz machine of Dayan *et al.* [8] was to follow in the footsteps of the Boltzmann machine [1] by regarding the two arrays of Ising spins as physical systems. From this perspective, minimization of the information cost of the descriptions of the states of the recognition and data generation is equivalent to minimizing the free energy of this physical system of interacting Ising spins. This in turn implies [8,9] that the conditional probability for a given model, i.e. the l.h.s of Eq. (1.1), will be given by Bayes's formula. We want to extend this scenario by replacing the Ising spins with the Riemann surface degrees of freedom introduced in Chapter 5.

One possibility [52] for going in this direction would be to replace the array of Ising spins with the Ashkin–Teller (AT) statistical model [144] for a 2D array of spins with two Ising spins per lattice site. It was discovered by Kadanoff and Brown [145] that if the 4-spin couplings are carefully chosen, then the energy functional for the model has a Gaussian form similar in form to the energy cost functions that naturally appear in Kohonen self-organization [13]. When the spins at each lattice site are allowed to interact, these AT models share with self-organizing networks [108] the crucial property that for critical values for the spin couplings the AT model will possess string-like excitations where the information regarding the state of the observer/controller or environment can be represented by the shape of a possibly topologically non-trivial 2D surface. The emergence here of "self-organization" means that the original AT spin degrees of freedom are effectively replaced with Θ-functions representing the shape of a Riemann surface. This has the pleasant consequence that both the recognition and generative networks of the Helmholtz machine can then be described as path integral representations of the Lax equation for the KdV equation.

The story line here is that we want to use the double path integral formulation of Feynman and Vernon [47] to represent the string degrees of freedom in the two Helmholtz machine arrays; one representing the forward evolution of an observer/controller which includes an estimation for the innovation, and the other to represent the backward evolution of the system or environment. Formally, this amounts to replacing Feynman's original path integral with a double path integral of the form (see Appendix E):

$$\int e^{-iS[x(t)]} Dx(t)$$

$$\rightarrow \iint e^{-i\{S[x(t)] - S[x'(t)]\}} F[x(t), x'(t)] Dx(t) Dx'(t)\}, \quad \text{(E.3)}$$

where the "influence function" $F(x, x')$ represents the effect of the interaction between two separate quantum systems on their joint evolution. The exact form for $F(x, x')$ depends on the details of the two interacting quantum systems, but in general one can write [47]

$$F[x(t)x'(t)] = \exp\left\{ -\int_0^T \int_0^t (x(t')\alpha(t,(t')) - x'(t')\alpha^*(t,t') \right.$$

$$\left. - (x(t') - x'(t)))\}dtdt' \right\}, \quad \text{(7.20)}$$

where $\alpha(t, t')$ is a complex function that plays somewhat the same role as the real valued autocorrelation function $A(t, t')$ for a random time signal. When the environment consists of an assembly of quantum oscillators that are linearly coupled to the coordinates $x(t)$ of the observer/controller, $\alpha(t, t')$ has the following form:

$$\alpha(t, t') = -\sum_{\omega_i} \frac{g_i^2}{\hbar^2} e^{-i\omega_i(t-t')}, \tag{7.21}$$

where the ω_is are the frequencies of the oscillators in the 2nd oscillator array making up the "environment". For a single harmonic oscillator and an environment consisting of oscillators with frequencies ω_i, where $\Delta_i = \hbar\omega_i$ is the level spacing, the exponential factor in Eq. (7.11) becomes

$$\exp\left\{-\sum_{ij} \frac{g_i^2 M\omega_0}{2h\Delta_i} \int_{t_j}^{t_{j+1}} dt \int_{t_j}^{t} ds (x(s)e^{-i\Delta_i(t-s)}\right.$$

$$\left. -x'(s)e^{i\Delta_i(t-s)})(x(t) - x'(t))\right\}$$

As described in Refs. [47,50], this expression can be used to exactly evaluate the effect of a quantum oscillator environment on an observer. However, we are mainly interested in the more sophisticated problem of modeling an interaction between an observer/controller with an environment that resembles the way sensory data are collected in the real world, e.g. as a self-organizing map.

Because of its exponential form of $F(x, x')$, the double path integral in E.3 can be written as

$$Z(w_i(\sigma)) = \sum_i \exp[i(S_{1i} - S_{2i} + \Delta S_{12})], \tag{7.22}$$

where actions S_1 and S_2 can be identified with the Nambu action [100] for an ensemble of free strings, which is the area of the Riemann surface. The total action describing the Feynman–Vernon dynamics of two arrays of interacting strings will be a sum of the free string actions S_1 and S_2 for the two arrays plus an interaction term ΔS_{12} describing the interaction between the Riemann surfaces in the two arrays of the Helmholtz machine. This ΔS_{12} contribution generalizes the nonlinear springs connecting the square and round points in Fig. 2.1. As an illustration of what sort of interaction might replace the nonlinear springs in the Durbin–Willshaw setup [1], ΔS_{12} might be assumed to have the form [146]

$$\Delta S_{12} = \iint d\sigma_1 d\sigma_2 [\det(g_{ab}^1)]^{1/2} [\det(g_{ab}^2)]^{1/2} G(\dot{x}^2 - \dot{y}^2), \quad (7.23)$$

where g_{ab} is the metric for a Riemann surface and $\dot{x}^2 - \dot{y}^2$ is the Lorentz invariant distance between a position on a Riemann surface representing the observer/controller and a position on a Riemann surface in representing environment. This in consistent with the Chern–Simons interaction that appeared in our treatment of the TSP. However, the path of the salesman is fixed, so there is no Riemann surface associated with the salesman's trajectory. On the other hand, it will turn out to be of considerable interest in the case of the Helmholtz machine to consider what happens when the degrees of freedom of either array are frozen in time. In fact, this takes us back to the Segal–Wilson and T–O descriptions of exact solutions of the KdV equation in terms of holomorphic functions on a Riemann surface.

Our elucidation of this follows some ideas of Chu [146]. Carrying out the integration over the string degrees of freedom for the environment array in the double path integral in (E.3) leads to an action function for each string in the observer/controller array of strings of the form

$$S' = \sum_j \int_0^{\tau_f} d\tau \int_0^\beta dy \, \exp\left\{ c_s^2 \left(\frac{dx_i}{d\tau}\right)^2 - \left(\frac{dx_i}{dy}\right)^2 + q_j(x_i[y, \tau]) \right\},$$

$$(7.24)$$

where c_s is the speed of sound along the string, $q_i(x)$ is an effective potential for the motion of the string, and the sum over j recognizes that there may be qualitatively different models for representing the input data (e.g. Riemann surfaces with different topologies). We see here how an inhomogeneous wave equation for a string moving in a certain potential obtained can be associated with every model for the environment. The crucial step to go from Eq. (7.15) to the KdV equation is to freeze the dynamics of the array of quantum strings representing the observer/controller, which then allows one to replace the quantum dynamics of an array of strings with probability distribution for the shapes of the frozen Riemann surfaces with a path integral description of the thermal state of solutions of the Schrodinger–Hill equation (that had its origins in the theory of Lagrange triangles [125]:

$$Z(\psi_0, \psi_\beta, \beta) = \sum_j \int \exp\left\{-\int_0^\beta \left[\left(\frac{d\psi}{dx}\right)^2 + q_j(\psi(y))\right] dy\right\} D[\psi(y)].$$

$$(7.25)$$

Although the stochastic structure of the KdV equation was not immediately recognized in the initial flurry of papers on exact solutions for the KdV equation (cf. [75]), it is now understood (see e.g. [147]) that the KdV equation is equivalent to an ensemble of Fokker–Planck equations where the drift terms can be identified with exact solutions of the classical KdV equation. From our perspective the Gaussian noise in this Fokker–Planck process can be regarded as a consequence of the quantum fluctuations inherent in a path integral description of the Lax equation for the KdV equation. In this setting the classical Mumford–Rissanen MDL principle for the Helmholtz machine corresponds to Rose optimization [34] for the Lax equation. Furthermore, it is clear the Rose optimum is attained when the entropy created by the wake–sleep algorithm is minimized. Thus, the information aspects of Bellman optimization arise in a natural way from the Fokker–Planck description of the Helmholtz machine during wake–sleep interludes as the MDL limit for the Helmholtz machine is approached. This is completely consistent with the conventional view [19] that the rate of change of the optimal Bellman function is partly due to a "drift term" (aka the reward function), and partly due to stochastic diffusion. If we go back to our quantum predecessor of this Fokker–Plank process, we can identify the reward functions $q_i(y, t)$ as the descendant of the interaction term ΔS_{12} in our quantum Helmholtz machine. This result offers the tantalizing prospect of being able to use a quantum model for the Helmholtz machine to solve complex optimal control and RL problems.

Because of the emergence of a probabilistic description for both the observer/controller and system/environment, the application of the wake–sleep algorithm to our quantum version of the Helmholtz machine can also be interpreted as a stochastic game [53] involving two players, where the "payoff" for each player is the information gathered regarding the history of the adversary. Stochastic games provide descriptions for adversarial conflicts in many real-life circumstances. For example, in the terminology of von Neumann [101] the drama illustrated in Fig. 1.1 is a two-person zero-sum game where the gain for the squirrel — his life — is matched by the bobcat's loss — a good meal. (We can reveal that a short time after the picture was taken the drama ended happily for the squirrel.) John von Neumann published separate discussions of game theory, the mathematical foundations of quantum theory, and computer models for the brain, and curiously his original paper on two-person zero-sum games (see [90]) appeared at about the same time as his treatise on the mathematical foundations of quantum mechanics [95]. Therefore, it is easy to imagine that von Neumann had in the back of his mind that these topics were somehow related. One of our ambitions is to fill in the gaps between von Neumann's published works.

As possibly a first step in this direction, our quantum model for the Helmholtz machine seems to provide an elegant explanation for why the strategies that "solve" two-person stochastic games [53] are probabilistic in the sense that on average no other strategy can produce a better outcome [105]. It is almost obvious that our quantum theory of the wake–sleep algorithm can provide such a model for these strategies, where the two-players are identified with the recognition and model generation networks of the Helmholtz machine, and the "payoff" is the information gathered by each player regarding the history of the other player. If one accepts that the goal of the wake–sleep algorithm is to maximize information gathered regarding the adversary, then it naturally follows that the innovations in this information that appear during wake–sleep interludes contribute to the optimization of the Bellman–Issacs function [53] that defines the optimal choices of strategies for both players in a stochastic game. In addition, von Neumann's minimax solution for two-player zero-sum games emerges because the equations describing the evolution of the observer/controller and system/environment from one wake–sleep cycle to the next are linear because the quantum dynamics is linear. The wake–sleep algorithm then forces these solutions to be compatible in the sense that their density matrices for the observer/controller and system/environment are similar, and the accumulated reward for the observer/controller is just the negative of the accumulated reward for the environment/system. That the expected equilibrium payoff for one player is just the negative of the expected equilibrium payoff for the other player is a consequence of the fact that the quantum dynamics for the observer/controller is forward propagating, while the quantum dynamics for the system/environment is backward propagating. (In the Feynman–Vernon formalism, this forward and backward propagation also applies to density matrices.)

A previous hint that optimal control might be related to von Neumann's solution for two-person games was provided by H_∞ control [84]. H_∞ control is an extension of the Kalman filter that constrains the control history in such a way that the variance of the innovation for the Kalman filter [83,84] remains bounded in magnitude. This type of constraint on the innovation also shows up in our quantum solution of the TSP problem [36] when we try to extract the optimal solution for the TSP by taking the classical limit. As noted previously, approaching the classical limit of quantum description of optimal control will in general require that the control histories must be describable as paths on a Riemann surface. Our focus here is that H_∞ control provides a nice way of looking at both the von Neumann and Nash solutions for games [105]. N-person games can be introduced by replacing the holomorphic functions in the Hardy space used in classical H_∞ control with the Gross–Pitaevskii description of the order parameter for N particles discussed in Section 7.1. In a Gross–Pitaevskii description for the "fluid states" of either the observer/controller or system/environment, each player represents a team of N agents that are treated symmetrically, and their dynamics will be controlled by the negative of a Hamiltonian like that in Eq. (7.16). The possible relevance of using a fluid-like model for an observer/controller is illustrated in Fig. 7.1, which shows a SOM model [12,13] for the somatosensory neurons on a human hand. As we shall see, SOMs may be a useful way to emulate H_∞ games.

As discussed in Section 4.2, the equilibrium state for an H_∞ game will be determined by maximizing a control Hamiltonian.

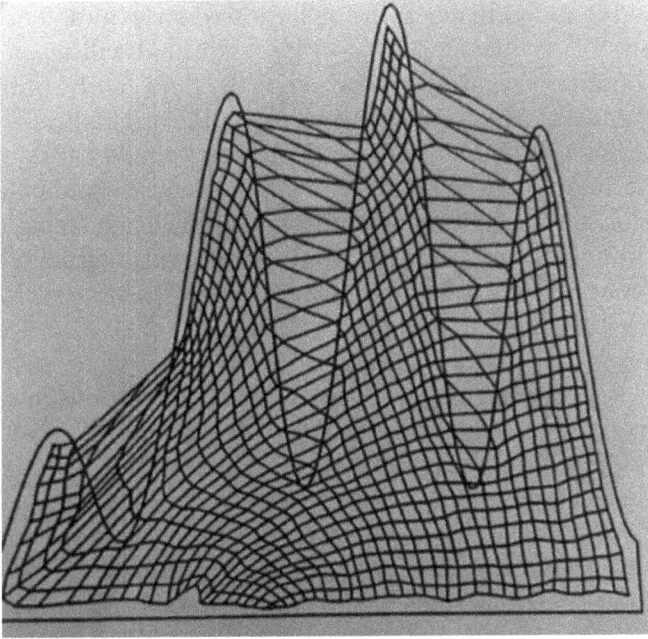

Fig. 7.1.   Self-organization of somatosensory sensors from Ref. [13].

On the other hand, the equilibrium state of either player can also be described as the state where the information that the team of agents representing a player has gathered regarding the state of the adversarial player is maximized. In the equilibrium state, the optimal strategies $\{\pi_i^*\}$ for the 2N agents representing the two players satisfy the Nash condition [105]:

$$e(\pi_1^*, \pi_2^*, ..\pi_i^*, ... \pi_{2N}^*) \geq e(\pi_1^*, \pi_2^*, ..\pi_i, ..., \pi_{2N}^*), \quad i = 1, ..., N,$$
(7.26)

where $e\left(\{\pi_i\}\right)$ is the game payoff for two teams of N agents representing the players in a two-player stochastic game. The $\{\pi_i\}$ are a set of mixed strategies for each agent, while the ith strategy on the r.h.s of the inequality (7.26) is any strategy other than the optimal strategy. The strategies with an asterisk are the optimal strategies that define the Nash equilibrium state. The Nash equilibrium condition in Eq. (7.26) is formally the same as optimization of the Bellman–Issacs function [53], which in this case sums up the payoffs for all the agents acting individually with the indicated strategies. The equilibrium state for a controlled Gross–Pitaevskii fluid will mean that no more information regarding the state of the fluid can be obtained by the independent actions of any agent. As in the original version of the Helmholtz machine due to Dayan, Hinton, *et al.*, [8] the equilibrium state of Gross–Piteavskii fluid can be accessed by minimizing the information gathered by the two players. A challenge facing this proposal is to find some means for implementing this regression. In this respect, construction of self-organizing maps with large scale parallel computers [147] may be the path to consider.

Although real games typically involve a discrete sequence of actions, while H_∞ control involves continuous variations in state and control variables, it might be useful to regard the H_∞ game as an approximation to the discrete sequence of actions for the multi-agent teams of agents. One reason this might be an interesting avenue to explore is that H_∞ control provides a natural home for Kohonen self-organization where both sensory data and explanations involve holomorphic functions. This connection with self-organization will be discussed further in the next section.

7.4.   Ad Mammalian Intelligence

Despite numerous advertisements regarding their successes and potential for artificial intelligence, neural networks have failed to live up to the initial hopes that they would provide definitive insights into why mammalian cognition works so well with a footprint dramatically smaller than the computational resources devoted to RL problems of interest (see e.g. [106]). Admittedly DNNs have provided many successes for machine learning, e.g. pattern recognition. On the other hand, even before the most impressive successes of DNNs appeared, a singular fundamental insight regarding mammalian cognition and artificial neural networks came to light in the form of the self-organizing maps for sensory data developed in the early 1980s by Teuvo Kohonen and his collaborators [12,13].

The formal theory of self-organizing maps is based on the notion of an output signal $w(r)$, which initially we will assume is a complex number describing a feature of its environment. We will posit that these "feature detectors" are located at an arbitrary set of points $\{r_i\}$ on a 2D surface, and evolve with time according to the following rule of form [12]:

$$w(r_j, t+1) = w(r_j, t) + \Lambda(r_j - r_{j*}) \, |\xi_j - w(r_j, t)| , \qquad (7.27)$$

where r_{i*} is the position of the neuron whose output is initially closest to an input feature ξ_j, while the function $\Lambda(\boldsymbol{r})$ leads to a "receptive field" for each detector which is the union of all input signals from a particular field of view producing a response in the detector at position r_i that is closer to the current state of the detector at r_i than the state of any other r detector. $\Lambda(|r_i - r_{i*}|)$ is typically a Gaussian function that allows the feature detectors to adjust their outputs so that not only the detector located at r_{i*}, but also nearby detectors observe the signal ξ_j. The "receptive field" for a sensor is the union of all input signals from a particular field of view that produce a response in the detector at position r_i that is closer to the current state of the detector at r_i than the state of any other r detector (cf. the receptive fields for the TSP defined by the dashed lines in Fig. 2.1). The self-organizing algorithm (7.27) adjusts the response of the sensor at position r_i to a particular environmental stimulus ξ_μ in its receptive field to be at least as strong as any of its nearest neighbors.

Of great importance for us is that Kohonen's self-organization maps also give rise to holomorphic functions that can serve mammalian cognition in much the same way that the holomorphic functions introduced in Chapters 5 and 6 can be melded together to provide an analytic model for the reward function for optimal control. In this regard, Ritter and Schulten have shown [13] that under the influence of random variables ξ_μ, the model outputs $\{w(r_i)\}$ evolve to a state which minimizes a stochastic energy functional

$$E[w(\vec{r}_j)] = \frac{1}{2} \left[\sum_{<r,s>} \sum_{\xi_\mu \in R} P(\xi_\mu) |\xi_\mu - w(\vec{r}_j)|^2 \right] . \qquad (7.28)$$

The reason that a sum over neighboring nodes appears on the r.h.s of Eq. (7.28) is that at each time step the change in position of a node affects its neighbors, which allows the entire ensemble to relax to a statistical configuration described by a partition function

$$Z = \sum_{L} \prod_{i=1}^{F} \oint dw(\vec{r}_j) \exp\left(-\frac{\kappa}{2} \sum_{i,j} \lfloor w(\vec{r}_i) - w(\vec{r}_j)\rfloor^2\right), \quad (7.29)$$

where the sum over L means a sum over triangulations covering the surfaces in $Z[w(\sigma)]$, each triangulation consisting of triangles F. Ritter and Schulten [13] show that in this equilibrium state the outputs $\{w(r_i)\}$ can be approximated by a continuous function satisfying the Cauchy–Riemann equations, which is the precise definition of a holomorphic function (cf. Appendix B). This connection of an SOM from the space of input signals to the space of models $w\{(\vec{r}_j)\}$ with smooth surfaces and holomorphic functions is undoubtedly the key to understanding how the analytic methods introduced in previous chapters relate to mammalian cognition.

If the input signals ξ_μ in Eq. (7.28) can be represented as a stationary random variable, a self-organized detector network will relax [13] to an asymptotic state characterized by a stationary probability distribution $P(\{w(r_i)\})$ for the various possible configurations of detector states, which in turn is derivable from a partition function of the form $Z = \exp(-E[w])$:

$$E[w] = \frac{1}{2} \sum_{i,j} C_{ij} |w_i - w_j|^2, \quad (7.30)$$

where C_{ij} is a covariance matrix for the detector states regarded as random variables. Fundamentally, looking for a pattern in the input data that is related to the underlying explanation for the data, Eq. (7.29) reveals [109] that this is equivalent to minimizing the area of a certain — possibly topologically nontrivial — surface, whose area will turn out to play the role of the information description cost. In addition, the energy functional (7.30) is the calling card for a holomorphic function on a surface. It seems to be a reasonable guess that these holomorphic functions play much the same role as the "Hardy spaces" which animate the RH solutions of both the KdV and NLS equations. Thus, we are invited to compare the results of self-organization with the exact analytic solutions to the integrable PDEs described in Chapter 5 and in T–O. In fact, Fig. 7.1, which shows a self-organizing map for the somatosensory sensors on a human hand, illustrates a dramatic result of such a comparison: the sensors come in different varieties; corresponding to the finger they are on. On the other hand, this association of different types of somatosensory sensors with fingers exactly mirrors the association of the number of solitons in a multi-soliton solution [76] of the KdV equation. This coincidence not only confirms our hypothesis that analytic solutions of integrable PDEs are important for understanding mammalian cognition, but suggests that the little understood reason for the leap in cognitive capabilities that is generally associated with the appearance of mammals is the ability to make use of various types of information.

This comparison between exact solutions of the KdV equation and SOMs makes sense of von Neumann's parallel interests in game theory, the mathematical foundations of quantum theory, and computer models for the brain. What von Neumann apparently didn't appreciate, or at least didn't put to paper, is the importance of analytic solutions of integrable PDEs as well as SOMs. These maps do in fact seem to have something in common with the way neurons are organized and process information in the cerebral cortex [107]. Furthermore, SOMs provide a "deer trail" heading in the direction of practical applications [109,110] for the holomorphic functions generated by self-organization. In a sense this is the most important technical result of our presentation because the spontaneous creation of holomorphic functions within the mammalian cerebral cortex might explain how it is that the mammalian brain can tap into the same analytic apparatus for generating reward functions and control actions extrapolated into the indefinite future based on solvability of PDEs. The bottom line is that Kohonen's self-organization provides what ultimately can be the most useful link between mammalian cognition and quantum mechanics.

Self-organizing networks have a property that they can organize sensory data in a way that one can understand what the data mean by "visual inspection". This is perhaps a realization of the often-quoted pearl of wisdom that a picture is worth a thousand words. This is illustrated in Fig. 7.1, which shows how an array of somatosensory sensors self-organize information about a 3D object so that a natural picture of the object is created within the sensor array itself. There is already evidence from MEG recordings [150] that different audio patterns are recorded in different areas of the cerebral cortex. The ability to provide a holistic understanding of different features of an environment is of course probably one of the reasons for the evolutionary success of mammals.

Chapter 8

Holistic Computing

8.1.   Quantum Mechanics and 3D Geometry

One nice feature of the Schwinger angular momentum representation is that measurement of an angular momentum state it might reveal much about the 3D shape of a quantum state as in the original Stern–Gerlach experiment. Indeed, the Schwinger representation can be used to both describe locations on the surface of a sphere with spherical harmonic function, but also encode whether the object being sought is present or not at the associated location by attaching a quantum qubit to any approximate location. For example, in the case of a Bayesian search for an object in an unknown location along a road, one may label the progress of a search by a sequence of qubit measurements. Ironically, in the context of these Schwinger states the essence of a Bayesian search would be a series of measurements *a la* the Stern–Gerlach experiment.

The role played by measurements in transforming Hilbert spaces into Hardy spaces memory into a universal calculator illustrates the more sophisticated role that can also can play in forming holistic images of an environment. Our teleportation scheme is a generalization of the scheme introduced in Ref. [4] for teleportation of a localized quantum state $|\psi(q_i)>$ represented as a function of a continuous variable q_i attached to a position i along a quantum wire represented as a one-dimensional graph of nodes. As an example, this variable might be the position of an oscillator located at i. Teleportation of this quantum state between two neighboring nodes i and $i+1$ of the quantum wire is accomplished by using a controlled phase gate, $\exp(iQ_{i+1} \otimes Q_i)$, where Q_i and Q_{j+1} are the position operators attached to the neighboring nodes, to entangle the quantum states $|\psi(q_i)>$ and $|p_{i+1} = 0>$ associated with these neighboring nodes, and then making measurements of the momentum operator P_i , where $[P_i, Q_i] = -i\hbar$, attached to the node i. In our scheme for teleportation of geometric objects the one-dimensional graph used in Ref. [4] to represent a quantum wire will first be replaced by a one-dimensional array of columns of nodes as illustrated in Fig. 1.1, where each node represents a two-dimensional quantum oscillator. The teleportation of three-dimensional objects will then be accomplished by generalizing the array of nodes shown in Fig. 3.1 to a three-dimensional array of nodes, where each node represents a four-dimensional oscillator.

We begin by generalizing our discussion of quantum teleportation of qubit states to teleportation of angular momentum states. In this case both the input data and underlying models can also be expressed in terms of eigenstates of the total angular momentum operator $J = L + S$ with values $j = l \pm 1/2$ and the z-component of the angular momentum J^z with values $j_z = m \pm 1/2$. The initial wave function for the hidden layer will be chosen to be a product of the $|\phi_i = 0>$ states for each node:

$$\Psi_H(t = 0) = \prod_i \sum_{j,j_z} |jj_z >_i) \tag{8.1}$$

The possibility of using quantum entanglement between quantum systems separated by some distance to teleport quantum information [1] has been discussed mainly in the context of teleporting quantum information represented by qubits. In this section, we note the possibility of using quantum entanglement of angular momentum states to teleport a model for a three-dimensional geometric object from one location to another. Our proposal extends previous proposals [39] for universal quantum computation using continuous variable quantum systems; and in particular the possibility of using quantum measurements to teleport quantum states of a continuous variable defined on the nodes of a graph. The particular continuous variables that carry information of our teleportation scheme are those associated with coupled quantized angular momentum states. As was first recognized by Wigner and Racah [112], the algebraic properties of angular momentum coupling coefficients for 3 or more angular momenta are intimately connected with the geometric properties of simplexes constructed from triangles. Honeycombs of tetrahedrons can also serve as models for three-dimensional objects with an arbitrary shape. Our scheme for the teleportation of a geometrical object depends on decorating the nodes of a three-dimensional graph with quantum angular momentum states. Teleportation is accomplished by introducing an entangling interaction between two neighboring triples of angular momenta states representing two faces of a tetrahedron sharing a common edge, and making measurements of the angular momentum components parallel to the shared edge.

In our two-dimensional generalization of the continuous variable teleportation scheme in [114], the quantum states attached to the nodes of the graph are assumed to be eigenstates of angular momentum states with a definite value for $J_i^2 = j_i(j_i + 1)$. The "position" operators Q_i attached to the nodes of a graph are the angular momentum operators J_i^z which generate rotations of the state attached to node i around a fixed axis (defined for each node), while the "momentum" operators P_i are the operators ϕ_i corresponding to the angle of rotation about this axis. The angle variables ϕ_i suffer from the well-known problem that they cannot be represented as self-adjoint operators; so we follow the familiar prescription of using instead $\sin \phi_i$ and $\cos \phi_i$ variables, which in our setting appear as the following displacement operators:

$$e^{i\phi_i} = \frac{J_i^x + iJ_i^y}{\sqrt{J_i^2 - J_i^z(J_i^z + 1)}}. \tag{8.2}$$

In our scheme the single quantum wire in Ref. [4] used to teleport wave functions of a continuous variable q is replaced by a two-dimensional graph for teleporting entangled states of three angular momenta. Quantum states representing triangles are created by entangling the angular momentum states attached to three neighboring nodes in a column of the quantum wire — as illustrated in Fig. 6.1 — so that the sum of the three angular momenta is zero; i.e. the angular momentum vectors form a perfect triangle in the classical limit. In a basis where the angular momentum component J^z along a fixed axis is well defined, the state representing a triangle is the sum of a product of the three states $|j_i m_i>$, where the coefficients are Wigner's $3j$ symbols [114]. In this chapter, we will focus on the possibilities of teleporting entangled graphical states associated with a compact subset of the entire graph which represent an array of triangles in two- or three-dimensions. We will assume that initially the quantum states attached to the nodes $\{i\}$ of the graph are always defined in a basis where the eigenvalues of the operators J_i^z have definite values. Motivated by the scheme described in Ref. [4] for the teleportation of continuous variable quantum states, we imagine teleporting a triangle along a one-dimensional path within the graph by combining the action of a controlled phase gate $\Pi_i \exp(iJ_{it}^z \otimes J_{i0}^z)$ acting between the angular momenta states attached to any column $\{i_0\}$ of the graph with a neighboring column $\{i_t\}$ with measurements of the angle operators $\sin \varphi_i$ and $\cos \varphi_i$ in the entangled state created by the controlled phase gate. In the following, we describe how these operations can be used to teleport geometric objects represented as quantum states.

Following the prescription in [113], the initial graphical wave function Ψ_{in} which describes the array of triangles that one wants to transport is a product of the entangled wave functions for triples of nodes within a set of "input" columns $\{i_0\}$:

$$\Psi_{in} = \Pi_{i \in \{i_0\}} \left(\sum_{m_a, m_b, m_c} \begin{pmatrix} j_{ia} & j_{ib} & j_{ic} \\ m_a & m_b & m_c \end{pmatrix} |j_{ia}m_a\rangle |j_{ib}m_b\rangle |j_{ic}m_c\rangle \right)$$

(8.3)

where the symbol in parenthesis is the Wigner $3j$ symbol. If the initial quantum state of the graph nodes outside the input columns is a product of angular momentum zero modes $|J_i^z = 0>$, teleportation of Ψ_{in} to neighboring "target" columns $\{i_t\}$ is accomplished by first using the controlled phase gate C_Z to entangle Ψ_{in} with the $|J_i^z = 0>$ states on the neighboring target nodes, and then measuring the momentum operators $P_i = \phi_i$ in the entangled state to effect the transformation:

$$\prod_{i_{ta}, i_{tb}, i_{tc}} \left| \sum_{J^z} J_{i_t}^Z \right\rangle \rightarrow \prod_{i_{0a}, i_{0b}, i_{0c}} X(\phi_{i_0}) R(\theta_{i_0}) F |\psi_{in}\rangle,$$

(8.4)

where F is the operator that switches the J_i^z basis to the ϕ_i basis, $X(<J_i^z>)$ is a shift operator that depends on the result of a ϕ_i measurement, and $R_z(\theta)$ is a rotation operator about common z-axis for the input and target column which can be applied either before or after the C_Z gate. The C_Z gate in Eq. (6.28) allows one to entangle an input state representing a triangle of angular momentum states with control states $|\Sigma J^z = 0>$ on neighboring columns, and then as a result of measurements of ϕ_i for the input nodes the wave function for the nodes in the next target layer $\{i_t\}$ become $X(\phi_{i0})\Psi_{in}(q_{i_1})$.

As an illustration of how quantum angular momentum states might be used for the teleportation of geometric objects, let us consider the teleportation of angular momentum states representing a triangle within a quantum circuit consisting of an array of two-dimensional oscillators. We envision that this array consists of 2D oscillators which are localized at points in three-dimensions and connected in a fashion at each node in the lattice is the used to define an angular momentum state in a basis where the J_i^z operators for all the nodes within each column have definite values with respect to a common z-axis. The wave function for the input layer corresponding to three nodes within the first column of the lattice has the form given in Eq. (8.3), where $i = 1, 2, 3$. The lines connecting the nodes correspond to controlled phase gates $C_Z = \exp(iJ_{it}^z \otimes J_{i0}^z)$ acting between nodes in neighboring columns. During the teleportation procedure, the z-axis used to define the angular momentum states for two neighboring columns of nodes is identified with the shared edge for two neighboring faces of a tetrahedron. The teleportation then rotates the quantum state representing one face about this edge so as to represent the neighboring face of a tetrahedron. In particular, if the rotation angles θ_i appearing in Eq. (8.4) are chosen to be the dihedral angle between neighboring faces of a tetrahedron, then by carrying out the dihedral rotations and measuring the momentum operators $J^x + iJ^y$ for the entanglement of the input state with the product of J^z zero modes for the target layer nodes, the input state is transformed into a quantum state for three angular momenta which represents a distortion of the face of a tetrahedron. By removing the distortion using the squeezing operator $X(m)$ and concatenating this process three times over, one can use the quantum circuit represented in Fig. 7.1 to construct all four faces of a regular tetrahedron.

Extending the quantum circuit represented in Fig. 8.1 by adding more columns of nodes would allow the basic teleportation step (2) to be repeated indefinitely; in effect allowing the quantum circuit to function as a "quantum wire" for triangles. Adding columns of nodes

Example: A double pyramid can be constructed by teleportation along a "figure 8" knot

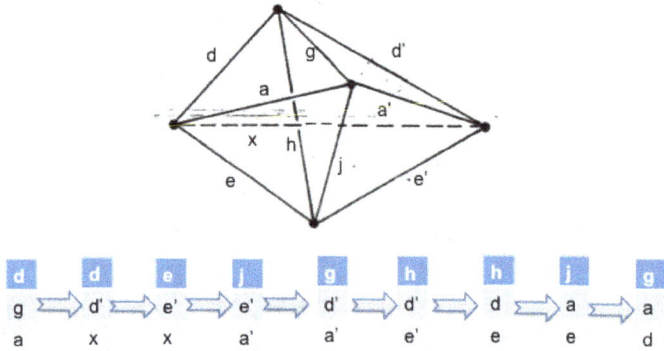

Fig. 8.1. Teleportation sequence illustrating the connection between quantum models for three-dimensional objects and knots.

along an axis perpendicular to this quantum wire would allow one to use our basic teleportation step, Eq. (8.4), to teleport a triangle to an arbitrary location in the resulting square lattice of columns. It might be noted at this point that extending teleportation along a straight wire to teleportation along an area filling a curve within the two-dimensional array of columns could be used to construct an area filling array of triangles. In this way construction of arbitrary two-dimensional shapes becomes possible, since the interior region of any closed planar curve can be approximated by a close-packed two-dimensional array of triangles. As Lagrange first pointed out [146], an equilateral triangle configuration for three bodies interacting via an inverse square force is quasi-stable. In the quantum case this quasi-stability of the classical configuration may translate to robustness of the Schwinger/Wigner representation of an equilateral triangle against quantum noise.

Generalizing the chain of clusters of 2D oscillators representing our Lagrange triangles to a two-dimension array of the clusters would allow a chain of tetrahedrons where neighboring tetrahedrons share a common edge. Because it is clearly possible for the chain of tetrahedrons to connect any two given points in the two-dimensional array of columns of triple nodes, we have a constructive proof that it is possible to teleport a tetrahedron between any two locations in a two-dimensional plane. This implies that there must be some subtlety involved in trying to generalize this scheme for constructing arrays of tetrahedrons to three-dimensions, because in general paths in three-dimensions are "knotted", which implies that a representation of the teleportation path lying within a two-dimensional array of columns of nodes must necessarily intersect itself.

The question of some interest to us is whether there is some generalization of the quantum circuit in Fig. 8.1 that would allow one to construct three-dimensional space-filling arrays of tetrahedrons at arbitrary locations in 3D. As just noted, one can teleport a two-dimensional closed curve by approximating the two-dimensional region inside the curve by an array of contiguous triangles. This suggests that one could approach the problem of teleporting three-dimensional shapes by first approximating the three-dimensional object by a space-filling array of tetrahedra. However, space-filling arrays of tetrahedra cannot in general be constructed using the two-dimensional teleportation circuit for triangles discussed above because a mapping of the triangle teleportation path for a three-dimensional array of tetrahedrons would be self-intersecting. The difficulties with using the triangle teleportation scheme illustrated in Fig. 8.1 can be understood by noting that the outer faces of the double pyramid can be constructed in a straightforward way using the scheme in Fig. 1.1; however, construction of the entire figure requires a "knotted" teleportation path.

One possibility to generalize the teleportation scheme illustrated in Fig. 8.1 so as to teleport tetrahedrons within a space-filling array of tetrahedrons would be to extend the Schwinger oscillator representation of the SU(2) angular momentum algebra to an oscillator representation of the Pauli $O(4) = SU(2) \times SU(2)$ algebra [10]. For the purposes of constructing quantum states which correspond to $O(4)$ representations, the one-dimension nodes of a three-dimensional tetragonal graph generalizing the 2D lattice in Fig. 8.1 can then be labeled by the eigenvalues of two independent angular momenta operators, J and K. Remarkably, if we utilize the freedom of labeling the quantum states attached to nodes with J and K quantum numbers, then the basic teleportation step Eq. (8.4) operating between nodes of a tetragonal three-dimensional graph will allow us to construct a finite space-filling array of tetrahedra. Since three-dimensional objects with a smooth boundary can be modeled as space-filling array of tetrahedra, the replacement of $SU(2)$ angular momentum states a tetragonal with the $O(4)$ of four-dimensional pentatopes analogous to a space-filling array of tetrahedrons. The pentatope is the four-dimensional generalization of a tetrahedron which contains 4 vertices, 10 edges, 10 triangles, and 5 tetrahedrons.

Remarkably, our approach to representing geometric figures is closely related to theories of quantum gravity in three-dimensions. In particular, tetrahedral simplexes and their relation to $6j$ symbols play an important role in a model for quantum gravity in 2+1 dimensions that was introduced some time ago by Ponzano and Regge [155]. The Ponzano and Regge model is distinguished in the sense that in the semi-classical limit the quantum action is related to a well-known topological invariant for three-dimensional knots, the Alexander polynomial. If the angular momentum states attached to the nodes in the figure are replaced by representations of the quantum group $SU_q(2)$, then it turns out that the effective quantum action for the tetrahedral network is related to the Jones theory of knots. In the semi-classical limit, the quantum action for Ponzano–Regge becomes the partition function for a classical statistical model. This brings us full circle back to statistical models, such as the quantum AT model discussed in Chapter 7, related to Bayes's formula Eq. (1.1).

8.2. Cognitive Science and Quantum Physics

We end our presentation with some musings about the deeper significance of the connection between quantum mechanics and Bayesian inference. It is easy to get the impression from the chaotic literature devoted to data science that there is nothing particularly mathematically profound about machine learning algorithms. On the other hand, one of our aims with this book is to frame the question of the mathematical significance of Bayesian inference in terms of its relationship to quantum mechanics. John von Neumann apparently thought that there was something mathematically profound about quantum mechanics. However, von Neumann did not clearly articulate what this means in his publications. Although our presentation has been focused on the relationship of the Bayes formula and quantum mechanics, we believe that our results may also shed light on two perennial philosophical puzzles: (1) What lies beneath the effectiveness of sophisticated mathematics for describing natural phenomena? As has been emphasized by Wigner [114], there is no apparent reason why this should be so, and (2) What is the meta-mathematical meaning of mathematics? This has been a puzzle since the time of Euclid and Plato.

If we adopt the utilitarian point of view that mathematics is simply an elaboration of the methods that the human brain has developed to solve practical problems [115], then at least to some extent the second puzzle can be absorbed into the more general question as to how the human brain works. Of course, whatever organizational features of the human brain are responsible for its capability to invent mathematical methods to solve practical problems, it seems reasonable to believe these features must have their origin in the way the cerebral cortex of primitive primates, e.g. lemurs, are organized. This in turn may mean that some simple organization principle such as Kohonen self-organization may have some relevance for understanding the emergence of mathematics. In view of the connections between Kohonen self-organization, holomorphic functions, and quantum mechanics that we have discussed in previous chapters, we can begin to see how the subjects of cognitive science, quantum mechanics, and the usefulness of mathematics may be intertwined.

It is interesting to note in this connection that although John von Neumann argued that digital computers might provide a model for how the human brain works, he was also apparently very interested in the relationship between quantum mechanics and mathematics. Indeed, it is perhaps not entirely coincidental that his most famous mathematical result, the minimax theorem for zero-sum games [117], was obtained at the same time (circa 1928) that he was working on quantum mechanics. In his bio submitted to the National Academy of Sciences (NAS), von revealed that "The part of my work I consider most essential is that on quantum mechanics, which was developed in Gottingen in 1926, and subsequently in Berlin in 1927–1929". Although to the author's knowledge von Neumann never suggested that quantum mechanics might be directly relevant to how the brain worked, it does seem that our results complement von Neumann's intuition regarding the joint importance of quantum mechanics and game theory. Apparently, what von Neumann missed is the usefulness of integrable dynamics and Kohonen self-organization for tying together quantum mechanics, game theory, and cognitive science. The ability to construct holistic three-dimensional images to represent data is an example of a capability which quantum devices seem to share with the mammalian cortex, and in both cases conveys a certain superiority over conventional computers.

Along these lines Kohonen drew attention to the usefulness of self-organization for analyzing observational data or robotic control [12], it was realized [13] that self-organizing maps also satisfy the Cauchy–Riemann equations. Evidently, at the heart of the remarkable capability of the mammalian brain to solve in real time Bayesian pattern recognition or decision problems that would be very time consuming — if not intractable — with conventional computational resources is the ability of the cerebral cortex to in effect solve the Cauchy–Riemann equations. On the other hand, efforts to emulate Kohonen self-organization using large-scale parallel computers [147] may be a first step to understanding how this capability might be approached with conventional computational resources. Further direct investigation of how brain wave patterns are related to cognition [154] may benefit [155] from the analytic techniques for solving Bayesian inference problems, like those discussed in earlier chapters, seems warranted.

Appendices

A. Gaussian Processes

A Gaussian process (GP) is a vector in an infinite dimensional space where the vector components are *iid* Gaussian random variables. GPs are universally useful in representing and analyzing data (see e.g. [148–149]). In the following, we will use the symbol D_P to denote the training data consisting of N pairs, $(\mathbf{X}_N, \mathbf{Z}_N)$, of input Gaussian processes $\{x_i^{(n)}\}$ and associated scalar labels $\{l^{(n)}\}$. In common with "supervised" neural networks, this approach to pattern recognition requires introducing examples of input data together with labels distinguishing different types of data. However, in contrast with artificial neural networks this approach does not necessarily require that an interpolating function $z(x)$ for the training data and data labels have a relatively simple representation in terms of known functions. Following the advice of MacKay [6], the best procedure for inferring the interpolating function $z(x)$ given a set of input data and training labels relies on the use of Bayes's formula Eq. (2.1):

$$P(z(x)|l_N, \mathbf{X}_N) = \frac{P(l|z(x), X_N)P(z(x))}{P(l_N, |X_N)}, \qquad \text{(A.1)}$$

When the training labels $l^{(n)}$ are real numbers, the problem of finding the function $y(x)$ is usually referred to as a regression problem. If $l^{(n)} = 0$ or 1, the inference problem is usually referred to as a search problem. Given a training dataset $D_N = \{\mathbf{X}_N, \mathbf{Z}_N\}$, the regression problem is to infer an interpolating function $z(x)$ which would allow one to infer the most likely target label l^* given a new input data vector $\{\mathbf{x}^*\}$. In general, it is not necessary to assume that the interpolating function $z(x)$ be parameterized in terms of a finite set of parameters; i.e. the dimensionality of $z(x)$ can be infinite. Thus, one might imagine that the computational complexity of finding $z(x)$ is in general prohibitive. However, since the input dataset is finite, it is generally sufficient to express $z(x)$ in terms of a set of basis functions that allow one to uniformly represent the features in the data represented by the labels $l^{(n)}$. It is often useful to assume that the signal $z(x)$ can be written in the form

$$z(x, w) = \sum_{m=1}^{m=M} w_m \phi_m(x), \tag{A.2}$$

where we will refer to the parameters w_m as weights. The number M of basis functions will of course depend on how accurately one would like to represent the distribution of data features represented by prior conditional probability for the labels $l^{(n)}$. The covariance matrix $C(z, z')$ for the signals $\{Z_N\}$ will be given by

$$C\left(z(x^{(n)}), z(x^{(n')})\right) = \sigma_w^2 \sum_m \phi_m(x^{(n)})\phi_h(x^{(n')}) + o_\nu^2 \delta_{nm'}. \tag{A.3}$$

In the ML schemes described by MacKay [6], the input data are in general characterized by a scalar label $\{l^{(n)}\}$ assigned to each specific example $\{\mathbf{x}^{(n)}\}$ of GP input data. If the labels are simply decimal numbers, then this data analysis task can be considered a "regression" problem, which is how we will refer to the task in the following. On the other hand, if the labels $l^{(n)}$ are discrete variables, then the task can be thought of as a "classification" problem. In the case of the Bayesian search discussed in Section 3.1, the obvious choice for this label is whether the object or state being sought lies near the location corresponding to the vector $\mathbf{x}_n \equiv \{x_1^{(n)}, \ldots, x_d^{(n)}\}$, where the superscript n corresponds to step n of the search. In general, the labels $\{l^{(n)}\}$ attached to GP data cannot be determined exactly, but with some error, and in general this error can be regarded as a Gaussian random variable. The methodology described in [6] involves constructing a continuous GP $z(\mathbf{x})$ to interpolate a set of labels $\{l^{(n)}\}$

for a training set $\boldsymbol{X}_N$ of input data $\{\boldsymbol{x}^{(k)}\}$ with say N examples (i.e. $k = 1, \ldots, N$), and then using this interpolation model to attach similar labels to new examples of data. MacKay's method assumes that each measurement $y(\boldsymbol{x}^{(k)})$ of $l^{(k)}$ differs from the model $z(\boldsymbol{x}^{(k)\}})$ by a random error:

$$y(\boldsymbol{x}^{(k)\}}) = z(\boldsymbol{x}^{(k)\}}) + v, \tag{A.4}$$

where v represents observational noise (It might be noted that Eq. (A.1) is an example of the general circumstance that the presence of some noise is always beneficial when it comes to fitting data). Using Bayes's formula, one finds

$$P(z(x)|Y_N, \boldsymbol{X}_N) = \frac{P(Y_N|z(x), X_N)P(z(x))}{P(Y_N, |X_N)}. \tag{A.5}$$

The power of using *GP*s lies in the ability to find analytic expressions for the conditional probability distributions $p(z(\mathbf{x})|\boldsymbol{Y}_N, \boldsymbol{X}_N)$ and $p(l_*|z(\mathbf{x}), X_N, x_*)$. Having these analytic expressions in hand allows one to quickly make probabilistic predictions for the label l_* that should be assigned to a new example of input data x_*. Using GP representations for $p(z(\mathbf{x})|\boldsymbol{Y}_N, \boldsymbol{X}_P)$ and $p(l_*|z(\mathbf{x}), X_N, x_*)$, the posterior conditional probability for finding label l_* at step N+1 in a Bayesian search can be found by integrating over all paths $y(\mathbf{x})$:

$$p(l_*|D, x_*) = \int dy(x)p(l_*|y(x), X_N, x_*)p(y(x)|D). \tag{A.6}$$

It is interesting that Eq. (A.3) provides a description of the evolution of a Bayesian search for the best interpretation of an observation as an integral over "paths", where the paths touch certain points as in the traveling salesman problem. This is a harbinger for the quantum schemes introduced in Chapter 7.

A new development in data analysis — the Gaussian process neural network (GPNN) [153] — offers an alternative to "wide" (i.e. many nodes in the hidden layers) neural networks that allows probabilistic pattern recognition to be carried out in a deterministic manner without the extensive training effort typically required to fix the connection strengths in deep back propagation neural networks. The GPNN has the elegant feature that the network outputs after L layers can be interpreted as the Bayesian prediction for obtaining an output Gaussian process z^* given an input Gaussian process x^*. In particular, given a set of input–output Gaussian processes $\{x^m, z^m\}$, where the pairs for $m = 1, \ldots, M$ represent the supervised training data, the Bayesian prediction for the output Gaussian process z^* in the case of a new input x^* is

$$P(z^*|D, x^*) = \int dz P(z^*|z, x, x^*) P(z|D) = \int dz P(z^*, z|x, x^*) \frac{P(t|z)}{P(t)},$$

(A.7)

where the "observation noise" $P(t|z)$ as well as the prior $P(t)$ are assumed to be normally distributed about z. If the variables x_i, z_i, and t_i are all assumed to be independent Gaussian distributed variables, the integral in Eq. (A.7) can be carried out analytically. The result in $z^*(x)$ is a Gaussian random variable with mean

$$\mu = K(x^* D)(K(D, D) + \sigma_\epsilon^2 I)^{-1} t$$

(A.8)

and variance

$$\sum = K(x^*, x^*) - K(x^*, D)(K(D, D) + \sigma_\epsilon^2 I)^{-1} t.$$

(A.9)

B.　Wiener–Hopf Methods

B.1.　*Cauchy–Riemann equations*

A complex function $f(p, q) = u + iv$ is *analytic* if u, v satisfy Beltrami equation:

$$\frac{\partial}{\partial q}\left[\frac{F \partial u/\partial p - E \partial u/\partial q}{EG - F^2}\right] = \frac{\partial}{\partial p}\left[\frac{F \partial u/\partial q - G \partial u/\partial p}{EG - F^2}\right],$$

(B.1)

where E, F, G are the metric coefficients for a smooth surface

$$ds^2 = E ds^2 + 2F dp dq + G dq^2.$$

(B.2)

If $F = 0$, then the Beltrami equation becomes the Laplace equation and u, v satisfy the Cauchy–Riemann equations:

$$\frac{\partial u}{\partial x} = \frac{\partial v}{\partial y} \quad \frac{\partial u}{\partial x} = -\frac{\partial v}{\partial y}. \tag{B.3}$$

The Cauchy–Riemann equations are the conditions that (u, v) represents a "holomorphic flow".

B.2. *N/D factorization*

A powerful addendum to the use of special functions to solve wave propagation problems was introduced in 1931 by Wiener and Hopf [61], who showed how sectionally holomorphic functions (functions that in some neighborhood of a chosen point of the complex pain can be written as a power series in z referring to that point as the origin) can be used to analytically solve scattering problems that would otherwise appear intractable. It is noteworthy from the point of view of this book that the Wiener–Hopf method was later applied with great effect to the problem of determining the potential in the Schrodinger from the asymptotic phases of a quantum wave function [60,117]. It is this later development which provides a crucial signpost for our quantum approach to machine learning. In the following, we will summarize how the Weiner–Hopf method makes much of the demonstration by Segal and Wilson [73] that the Riemann–Hilbert method for representing sectionally holomorphic functions allows one to construct a map between the two BFS spaces that is equivalent to the construction of the map between two Hilbert spaces that is the essence of essentially all machine learning algorithms. Indeed, an abbreviated summary of what we hope to accomplish in this book is that we will tie together the threads provided by the use of Riemann–Hilbert, Wiener–Hopf, and Segal–Wilson methods for analyzing integrable differential equations in order to provide a Hamiltonian framework for solving signal analysis, optimal control, and reinforcement learning problems.

The person who turned the Wiener–Hopf idea of using the theory of complex variables in order to provide exact descriptions of scattering problems discovery toward machine learning was Wiener himself. Working at the onset of World War II on the problem of extracting a signal from a time series of measurements contaminated with noise, Wiener's original derivation of his filter [81] involved solving an integral equation of the same form as the integral equation for wave scattering discovered by Wiener and Hopf. Not only was Wiener's discovery applied with good effect during the war, but in the years following WWII Kalman modified Wiener's signal filter in such a way as to address the problem of optimal control [83]. It turns out [41] that the mathematical structure underlying both the Wiener and Kalman filters involves functions that are rational functions (i.e. ratios of polynomials) of the wave frequency regarded as a complex number. Perhaps the most momentous aspect of the effort to find an exact solution of the KdV equation is that a certain rational function of the eigenvalue of the 1D Schrodinger operator regarded as a complex variable plays much the same role as the rational functions in the Wiener and Kalman filters. Thus, the parts of machine learning that flow from the work of Wiener and Kalman seem to involve in essence the construction of a certain rational function of a complex variable. It was left to a group of Russian mathematicians [119] as well as Segal [73] to point out [94] the connection of this effort with algebraic geometry and that the setting for this rational function is a Riemann surface rather than the usual complex plane.

A potentially very profound advantage of using quantum amplitudes rather than real probabilities to solve pattern recognition and decision tree problems arises from the observation that, in contrast with classical probability densities, the quantum amplitudes used to describe the state and evolution of a quantum system are always complex valued quantities that are typically analytic functions of the continuous variables describing the system. This allows one to take advantage of powerful methods for representing analytic functions in terms of their singularities in the complex plane. In particular, in 1931 Weiner and Hopf [61] made the remarkable observation that certain kinds of integral equations that arise in scattering problems can be solved by regarding the scattering amplitudes as analytic functions of the parameters of the problem. For example, it was shown in the 1950s that certain interesting problems involving the scattering of electromagnetic waves that one might guess are intractable, e.g. the scattering of an electromagnetic wave from a flat conductor with a knife edge, could be easily solved by extension of the physical solution to a solution where the frequency is a complex variable. In the case of the scattering of a quantum particle from a localized potential, the Weiner–Hopf method can be illustrated by considering the low energy scattering of a quantum particle. The asymptotic wave function is proportional to

$$u(k) = e^{ikr} e^{2i\delta} e^{-ikr}/r. \tag{B.4}$$

where δ is the S-wave phase shift. The scattering amplitude $f(k)$ is

$$f(k) = \frac{e^{i\delta}\sin\delta}{k}. \tag{B.5}$$

The time reversal symmetry of the Schrodinger equation implies that

$$f^*(k) = f(-k^*)$$

and $\tag{B.6}$

$$f^*(k^2) = f(k^{2*}).$$

These properties of the scattering amplitude are consistent with assuming that the only singularities in $f(k)$ regarded as a complex function of k^2 are branch cuts running along the real axis with an imaginary discontinuity across the branch cut. This allows one to use Cauchy's theorem to write $f(k)$ in the form

$$f(k^2) = \frac{1}{\pi} \int_{-\infty}^{-k_0} \frac{Im f(k'^2)dk'^2}{k'^2 - k^2 - i\epsilon} + \frac{1}{\pi} \int_0^\infty \frac{Im f(k'^2)dk'^2}{k'^2 - k^2 - i\epsilon}. \tag{B.7}$$

If we now write $f(k) \equiv N/D$, where N has singularities only in the left-hand complex plane and D has singularities only in the right-hand plane, then the Cauchy representation (C.8) becomes [152]

$$N(k) = \frac{1}{\pi} \int_{-\infty}^{-k_0} \frac{D(k'^2)Im f(k'^2)dk'^2}{k'^2 - k^2}$$

$$\tag{B.8}$$

$$D(k) = 1 - \frac{k^2}{\pi} \int_0^\infty \frac{N(k'^2)k'dk'^2}{k'^2 - k^2}.$$

The quantities $N(k)$ and $D(k)$ carry all the information regarding the potential that is necessary to construct the S-wave scattering amplitude for a particle as function of the momentum of the particle, and provide an elegant and unique path to defining the kernel function that appears in both Dyson's adaptive optics setup and Bayesian pattern recognition. In the context of quantum scattering theory the function $D(k)$ is known as the Jost function. This function plays an important role in the theory of inverse scattering in both one- and three-dimensions. In the case of the three-dimensions, $D(k)$ becomes a matrix [120], and the inverse scattering equations are known as the Newton–Jost equations.

B.3. *The Gelfand–Levitan–Marčenko (GLM) equation*

The problem of finding the potential of the 1D Schrodinger equation using the frequency dependence of the reflection and transmission coefficients for the potential (which is normally assumed to be spatially compact). This problem was first addressed by Marčenko [60]. In his 1955 paper, Marčenko showed that the Wiener–Hopf method used by Wiener to construct his noise filter could be used to solve the problem of finding the potential for the Schrodinger equation based on scattering data. A similar method was also developed by Gelfand and Levitan which relates causal and anti-causal Green's functions for the inhomogeneous wave equation. In order to explain how this works, we will follow Tao's nice derivation in his tribute to Israel Gelfand [59]. The GLM equations allow to infer the potential $u(x)$ that occurs in the inhomogeneous wave equation:

$$\frac{\partial^2 \psi}{\partial t^2} + \frac{\partial^2 \psi}{\partial x^2} = u(x)\psi(x,t) \tag{B.9}$$

based on scattering data, i.e. the obtained by measuring the reflection coefficient $R(k)$ as a function of wave number. If $u(x)$ were 0, then the solution to Eq. (B.9) would be a superposition of a right-moving wave $f(x - t)$ and a left-moving wave $g(x + t)$. If $V(x)$ is localized near the origin inside a finite interval $[-L, L]$, then the solution in the region $x < -L$ has the form $f_-(x - t) + g_-(x + t)$, while the solution in the region $x > L$ has the form $f_+(x - t) + g_+(x + t)$. Two special cases of these general solutions are of particular interest: (1) for large negative values of $t f_- = \delta(x - t)$ and $g_+ = 0$, while for large values of t the solution for $x < -L$ has the form $G(x, t) = \delta(t - x) + R(t + x)$

where $R(t + x)$ is the amplitude of the wave generated by reflection of the input pulse from the potential, and (2) the roles of x and t are reversed, so that as a result of a time-dependent potential a solution where $f_- = \delta(x - t)$ and $g_- = 0$ for $x < -L$ evolves to a solution of the form $u(x, t) = \delta(x - t) + K(t - x)$, where $K(x, t) = 0$ unless $x > t$. The Fourier transform of the function R is referred to as the scattering data while the Fourier transform of the matrix K is often referred to as the Jost function [51]. Multiplying the solution $\delta(x + t - s) + K(s - t, x)$ by $R(s)$ and integrating over s one can show that $\delta(x - t) + R(x + t) + \int K(x, s)R(s + t) ds$ is also a solution. Using the fact that $G(x, t)$ vanishes for $x > t$, one arrives at the integral equation:

$$R(t + x) + K(t - x) + \int_{-\infty}^{x} R(s + x)\, K s - x)\, ds = 0. \qquad (B.10)$$

Given the input data $R(t+x)$ the acausal covariance function $K(x, y)$ can be determined by solving the liner integral Eq. (2.15). This is the covariance function that is used in least squares stochastic estimation. The potential that appears in the wave equation (B.9) is given by

$$u(x) = 2\frac{d}{dx}K(x, x). \qquad (B.11)$$

The GL equations pertain to the scattering solutions of the time-independent form of Eq. (B.9):

$$\frac{\partial^2 u}{\partial x^2} - u(x)\psi(t, x) = k^2\psi(t, x), \qquad (B.12)$$

where $uv(x)$ is assumed to be everywhere positive ($u(x) > 0$) with compact support centered on the origin. The solutions to Eq. (B.12) that are of interest in connection with the inverse scattering problem are solutions which for $x \to -\infty$ have the form

$$\psi(k, x) = e^{-ikx} + R(k)e^{ikx}, \qquad (B.13)$$

when $R(k)$ is referred to as the reflection coefficient. For $x \to \infty$, the solutions of interest have the form

$$\psi(k, x) = T(k)\psi(k, x), \qquad (B.14)$$

where the reflection and transmission coefficients satisfy

$$|T(k)|^2 + |R(k)|^2 = 1. \tag{B.15}$$

The inverse problem is given $R(k)$ for $0 < x < \infty$ determine $v(x)$. The Marčhenko method [52] solves the integral equation

$$u(s, x) = R(s + x) + \int_{-\infty}^{t} R(\tau) u(\tau, x) d\tau, \tag{B.16}$$

where $s < x$. The potential for the inhomogeneous wave equation is

$$v(x) = 2 \frac{\partial}{\partial x} u(x, x). \tag{B.17}$$

Alternatively, the kernel $u(x, x)$ can be written as

$$u(x, x) = \frac{1}{2} \int_{-\infty}^{x} v(x) dx \tag{B.18}$$

As noted in the text, kernel functions $u(x, y)$ that satisfy (B.15) also arise in connection with finding least square estimators for signal filters and data feature predictors. In both these cases the interpretation of Eq. (B.11) as the time-independent Schrodinger equation is very useful as a guide to what sorts of kernels might be of interest. Indeed, the eigenfunction expansion considered in the original paper of Machenko *et al.* is a natural quantum Hilbert space representation for data features.

B.4. *The Riemann–Hilbert problem*

First posed by Riemann in his 1852 PhD thesis, the question of how to construct a complex valued analytic function from knowledge of its real and imaginary parts on a curve in the complex plane has been. In 1905, David Hilbert reshaped Riemann's problem into the solution of a certain nonlinear integral equation [127]. This equation is very similar to the integral equation introduced in the 1930s by Norbert Wiener and Eberhard Hopf for the scattering of electromagnetic waves [61]. Historically, the mathematical underpinnings of the Weiner–Hopf method go back to Hilbert's 1905 paper in which he considered a boundary value problem for a holomorphic complex valued function Φ_+ defined in a connected region S_+ of the complex plane bounded by a closed contour L:

$$\Phi_+ = G(t)\Phi_- + g(t), \tag{B.19}$$

where $\Phi_{\pm}\pm$ are the boundary values of holomorphic functions on S_+ and its compliment S_-, and $G(t)$ and $g(t)$ are smooth bounded functions on L. In the 1930s, it was shown that writing

$$G(t) = \frac{X^+(t)}{X^-(t)} \tag{B.20}$$

the general solution to Eq. (B.19) is [67]

$$\Phi(z) = \frac{X(z)}{2\pi i} \oint \frac{g(t)dt}{X^+(t)(t-z)} + X(z)P(z), \tag{B.21}$$

where $X(t)$ is the solution to Eq. (C.1) when $g(t) = 0$, and $P(z)$ is a polynomial. It might be noted that the integral in Eq. (B.3) can be expanded in inverse powers of z. This property plays a role in our discussion of the Baker function in Chapter 5.

As a result of the flurry of activity in the 1920s and 1930s, finding solutions of the Schrodinger equation for physically interesting problems, it was discovered that certain special analytic functions that had been discovered in the 19th century, were again found to be quite useful. Of special interest to us is that while he was a post-doctoral student at the Bohr Institute in Copenhagen, Lev Landau discovered the "special" analytic functions discovered in the 19th century which turned out to be of particular importance for quantum mechanics because they solve the Schrodinger equation for problems of great interest. Prominent examples are the hypergeometric functions. In the case of the Airy function, the problem is a particle subject to a linear potential; e.g. the potential acting on a particle near the surface of the Earth due to gravity. In fact, Airy functions are of importance for quantum motion in any potential because they describe the motion of the particle near the classical turning point.

A powerful addendum to the use of special functions to analytically solve linear differential equation was introduced in 1931 by Wiener and Hopf [61], who showed how sectionally holomorphic functions (functions that in some region of the complex plane can be written as a power series in z referring to some point in the region as the origin) can be used to analytically solve scattering problems that would otherwise appear intractable. It is noteworthy from the point of view of this book that the Wiener–Hopf method was later applied with great effect to the problem of determining the potential in the Schrodinger from the asymptotic phases of a quantum wave function [60,117]. It is this later development which provides a crucial signpost for our path to using quantum mechanics for machine learning. We will make much of the demonstration by Segal and Wilson [75] that the Riemann–Hilbert method for representing sectionally holomorphic functions allows one to construct a map between the two BFS spaces that is equivalent to the construction of the map between two Hilbert spaces that is the essence of essentially all machine learning algorithms. Indeed, an abbreviated summary of what we hope to accomplish in this book is that we will tie together the threads provided by the use of Riemann–Hilbert, Wiener–Hopf, and Segal–Wilson methods for analyzing integrable differential equations in order to provide a Hamiltonian framework for solving signal analysis, optimal control, and reinforcement learning problems.

$$\Psi(\xi, k) = \begin{pmatrix} 1 & \Gamma_k \int \dfrac{\exp\left(i\xi v + \dfrac{v^3}{3}\right)}{v - u} dv \\ 0 & 1 \end{pmatrix}, \tag{B.22}$$

which allows one to extract the solution for the homogeneous Painlevé equation as the residue of the Ψ_{12} matrix element as $\lambda \to \infty$. Its [68] calls this the "nonabelian Airy problem". In general, one can extract the integrand $g(\lambda)$ of an integral $\int g(\lambda)d\lambda$ over a contour by taking the limit $\lambda \to \infty$ of

$$Z(\lambda) = \begin{pmatrix} 1 & \int \dfrac{g(\mu)}{\mu - \lambda} d\mu \\ 0 & 1 \end{pmatrix}, \tag{B.23}$$

which yields the Wiener–Hopf factor

$$G(\lambda, x) = \begin{pmatrix} 1 & 2\pi i g(\lambda) \\ 0 & 1 \end{pmatrix}, \tag{B.24}$$

The Baker wave function is

$$\Psi(\lambda, x) = Z(\lambda, x)\exp\left\{ \left(\frac{4}{3}\lambda^3 + x\lambda \right) \sigma_3 \right\}, \tag{B.25}$$

where σ_3 is the Pauli matrix.

A similar approach works for the KdV equation using the real line as the contour [71]. $G(z, x)$ contains the scattering data for the Baker function, and the matrix linking the two holomorphic Hilbert spaces is

$$G(\lambda, x) = \begin{pmatrix} 1 - |r|^2 & re^{-2i(z^3t+xz)} \\ re^{2i(z^3t+xz)} & 1 \end{pmatrix} \Psi^-. \tag{B.26}$$

We now come to the punch line; this matrix contains all the information necessary to construct the Bellman and reward functions for the Kalman filter:

$$\Phi(z) = \frac{1}{2\pi i} \oint \frac{\Phi^+ - \Phi^-}{s - z} ds, \tag{B.27}$$

where

$$\Phi^+(s) = \Phi^-(s)G(s),$$

is a finite dimensional matrix equation. If the contour in (B.27) is a polygon, then the logarithmic jump matrix $G(s) = M_1 M_2 \dots M_k$ is a product of piecewise constant finite matrices. The rational function $\phi(x, \lambda)$ can then be constructed as a product of τ-functions:

$$\tau_k(Y) = Y(\lambda)M_k. \tag{B.28}$$

The appearance of a τ-function here is a strong hint that stochastic processes underlie the KdV dynamics [72].

B.5. *Inverse scattering transform*

The problem of determining the nature of an inhomogeneous medium from scattering data — the so-called inverse scattering problem — is of great importance in several contexts; e.g. geophysical exploration. Although in general these problems can only be approached numerically, the scattering of a plane wave by a compact inhomogeneity in three-dimensions is one example that is tractable because of the Wiener–Hopf method [61]. Newton [117] derived equations generalizing the GLM equations which can be used to infer the nature of a localized potential in three-dimensions from observations of the scattering of waves off the potential. As in the 1D case, the initial step was to introduce the Fourier transform of the scattering amplitude. The problem of determining the nature of an inhomogeneous medium from scattering data is of great importance in several contexts, e.g. seismology. Unfortunately, in general these problems can only be approached numerically.

Our quantum approach to Bayesian inference is based on an optimization principle for classical inverse scattering discovered by Rose [34]. The origin of Rose's principle is a method for solving the GL equation due to Marčhenko [60,117]. As in our discussion of the GLM equation in Section B.1.3, Rose starts by considering the solution $u_+(t, x)$ for the inhomogeneous wave equation [34,116] when the initial state is a delta function shaped wave incident from the left:

$$u_+(t, x) = \delta(t - x) + R(t + x), \quad \text{for } x < 0 \qquad \text{(B.29)}$$

and

$$u_+(t, x) = T(t - x), \quad \text{for } x < 0. \qquad \text{(B.30)}$$

The existence of the solution $u_+(x, t)$ allows one to write the solution to (B.7) when the incident wave $u_0(x, t)$ has any shape with a sharp wave front, i.e. $u_0(x, t) = 0$ when $x - t > x_0$, in the Green's function form:

$$u(t, x, x_0) = \int_{-\infty}^{\infty} u_+(t - \tau, x) u_0(\tau, 0, x_0) d\tau. \qquad \text{(B.31)}$$

The analog of the orthogonality condition for representing kernel eigen functions is

$$\int_{-\infty}^{\infty} u_+(t, x) u_+(t', x) dx = \delta(t - t') + R(t + t'). \qquad \text{(B.32)}$$

The usual orthonormality conditions for the time-independent Schrodinger equation can be recovered by Fourier transforming Eq. (B.34) with respect to both t and t'. Causality and the finite velocity of propagation imply that the scattered field (B.17) near the wave front is given by

$$u(t = 0, x_0^-; x_0) = u_+(t = 0, x_0^-; x_0) - \frac{1}{2} \int_{-\infty}^{x_0} v(x) dx.$$

The Rose optimization principle [34] is that the "best" choice for $v(x)$ is the one that leads to a scattering state that is entirely focused on a particular location x^* at a chosen time t^* in the future.

$$B = \int_{-\infty}^{\infty} [u(t, x; x_0) - \delta(t - x + x_0)]^2 dx = 0. \tag{B.33}$$

B.6. *Wave propagation with flexible boundaries*

In *Methods of Theoretical Physics* [116], Morse and Feshbach discussed in some detail how to calculate the motion of a string, a membrane, or elastic medium to an arbitrary force $f(t)$ applied to some limited region inside the medium or on its boundary. described a general for describing the propagation of sound waves in an elastic medium with a flexible boundary with its own dynamics. For an elastic medium that can be modeled as a string with length l, their approach was to write the solution as a double integral

$$\psi(x|t) = \int_0^l dx_0 \int_{-\infty}^t g(x|x_0||t - \tau) f(x_0|\tau) e^{-i\omega t} d\tau, \tag{B.34}$$

where it should be noted that the second integral only goes over times prior to t. They showed that problems of this type can be solved by first calculating the wave response $G(x, x_0, \omega)$ when a periodic impulse is applied at a particular point x_0 on the boundary of the medium carrying the wave:

$$\psi(x|x_0|t) = \int_{-\infty}^{\infty} G(x, x_0|\omega) F(\omega) e^{-i\omega t} d\omega, \tag{B.35}$$

where

$$F(\omega) = \int_{-\infty}^{\infty} f(t) e^{i\omega t} d\omega. \tag{B.36}$$

Unfortunately, when G as a function of ω has singularities on the real axis, then the integral in Eq. (B.35) is not convergent, and one must resort to considering ω as a complex variable. This led Morse and Feshbach [116] to consider replacing Eq. (B.36) with the Laplace transform:

$$F(s) = \int_{-0}^{\infty} f(t)e^{-st}d\omega. \tag{B.37}$$

Solving the wave motion with a flexible boundary, taking into account the distributed impulse resulting from the entire flexible boundary, can now be found by finding the "filter" function $g(x, x_0, t)$ for the coupled wave medium/flexible boundary whose Laplace transform is Green's function $G(x, x_0, \omega)$. Given this filter function, the complete motion of the string due to the imposition of a distributed force $f(x_0|t)$ along the length of the string can be obtained. We can do no better summarizing the Morse–Feshbach proposal for how to solve this type of problem than simply quote their synopsis in *Methods of Theoretical Physics* [116]:

"First compute the Green's function $G(x, x_0|\omega)$ for the steady state response of the system to a force of unit amplitude and frequency ω applied to point (x_0, y_0, z_0) within or on the boundary, by solving an inhomogeneous Helmholtz equation with an inhomogeneous boundary condition. Find the impulse function $g(x, x_0, t)$ for which G is the Laplace transform, either by contour integration of (Eq. (B.35)) or inverting (Eq. (B.37)). The response to $f(x, t)$ is then given by (Eq. (B.35))".

One beautiful feature of the Morse–Feshbach prescription for dealing with rubber potentials is that it not only illustrates the role of causality more clearly, but it also immediately illustrates why the nonrelativistic Schrodinger equation is relevant for understanding wave propagation with rubber potentials.

B.7.   *Adaptive optics*

A very interesting practical application of rubber potentials came about in 1970s because of efforts of [38] to remove the degradation of the angular resolution of ground-based astronomical telescopes due to atmospheric turbulence. In his 1975 paper on adaptive optics for reflective telescopes in the presence of photon noise, Dyson showed [58] that it is possible to remove the noise in an optical signal passing through the atmosphere due to turbulence by adjusting the shape of a reflecting mirror. The original approach to adaptive optics [38] made use of neural networks similar to those mentioned in Appendix A.

In the context of the first approaches to adaptive optics, these neural networks used as inputs the output of a phase sensor which detected deviations in wave fronts from flatness, and then used a feedback loop to adjust the shape of the mirror. The setup considered by Dyson similarly used a phase sensor's assessment of the shape of a flexible mirror surface Σ and then used a feedback circuit to mechanically deform the surface so as to compensate for small variations $a(\sigma, t)$ in the optical path length of light rays incident on a surface at various locations σ. Dyson proceeded by writing down equations describing the interplay between the controlled deformations in the shape of a surface and changes in the output of phase sensors which record changes in the intensity of light beams due to changes in the shape of the surface. These equations involve two matrices, $A_j(\sigma, t)$ and $B_j(\sigma, t)$. The matrix B supposes that we have a control system that adjusts the displacement $\delta(\sigma, t)$ of the surface with sufficient accuracy so that the signal intensity at time t and position σ is a linear function of the observed phase $\phi(\sigma, t) \equiv \delta(\sigma, t) + a(\sigma, t)$

$$I(j, t) = I_0 - \int d^2\sigma B_j(\sigma, t)\varphi(\sigma t), \qquad (B.38)$$

where $I(j, t)$ is the signal recorded at time t, in detector j and the $I_0(j)$ are the recorded signals in the sensor array in the absence of imposed variations in signal intensity due to atmospheric noise. The second equation relates the deformation of the surface at location σ produced by the feedback control to the observed signal intensity in the sensor array:

$$\delta(\sigma, t) = \sum_j \int_{-\infty}^{t} dt' A_j(\sigma, t - t')I(jt'), \qquad (B.39)$$

where the integral over d Ω means sampling the light intensity at a sufficiently large number of points on the sensor array as is required to determine the parameters which define the shape of the surface. When photon noise N is neglected, then Eqs. (B.1–2) have the classical solution (in matrix shorthand for A_j and B_j)

$$\varphi(N = 0) = [1 - AB]^{-1}a. \qquad (B.40)$$

Equations (B.38–B.39) define Dyson's adaptive optics model. What is most remarkable about his model though is that when photon noise is considered, the problem of changing the shape of the deformable mirror to compensate for random changes in the optical path of the illuminating beam across the mirror aperture becomes equivalent to solving the inversion problem for the multi-channel Schrodinger equation. This situation is qualitatively different from the classical case because in the classical case the negative feedback would amplify the photon noise. Dyson showed that in the presence of photon noise the optimal feedback matrices $A(\sigma, x, t')$ and $B(\sigma, x)$ remain finite and satisfy $A = KB^T I_0^{-1}$, where $K(\sigma_1, \sigma_2, t_1, t_2)$ is a matrix satisfying the nonlinear integral equation:

$$K + K^T + K(B^T I_0^{-1} B)K^T + U = 0, \qquad (B.41)$$

where U is the average $< a_1 a_2 >$ over time. Dyson showed that the solution to Eq. (B.41) is optimal in the sense that a quadratic function of the errors is minimized. Dyson also noted that Eq. (B.41) has the same form as the multi-channel Newton–Jost equation [117] that solves the inverse scattering problem for the quantum mechanical scattering of a quantum particle by a possibly anisotropic 3D potential.

C.   Riemann Surfaces

A Riemann surface is a topologically nontrivial smooth curved 2D surface whose points can be parameterized by the solution of an algebraic equation. The appearance of this Riemann surface is a result of a surprising theorem from the 1920s due to Burchnall and Chaundry [77] that the complex valued eigenvalues, say y and z, for two commuting differential operators constructed from the infinity of Lax operators parameterize a 2D surface defined by an algebraic equation of the form

$$y = \pm\sqrt{a_0 + a_1 z + a_2 z^2 + \ldots a_n z^n}, \qquad (C.1)$$

Riemann's greatest achievement was to point out that the ambiguity represented by the $\pm$ sign in an algebraic expression like Eq. (C.1) can be removed by the assertion that the solutions to (C.1) correspond to points on a topologically nontrivial 2D surface. The value of n in Eq. (C.1) becomes the "genus" g of this surface; i.e. the number of "handles" formed by the surface (cf. [55]). There is a complex torus, known in the literature [54,55] as the "Jacobian variety" of the surface, associated in a canonical way with any Riemann surface. In the case of the KdV equation, the Poisson–Arnold tori [91] emerges as the Jacobian variety associated with the Burchnall–Chaundry surface [77].

Algebraic geometry enters when for each torus one considers all possible pairings of the eigenvalues λ of the Lax operator $L(u)$ with the eigenvalues of any operator Q that commutes with $L(u)$. This is a complex curve C which can be mapped into a complex torus $\mathbb{C}^g/L$, where L is a g-dimensional complex lattice, by the Abel map $A(w) = \int_{w_0}^{w} d\omega$, when the integration path is a nontrivial 1-cycle on the surface.

$$A(w) = \int_{w_0}^{w} d\omega,$$

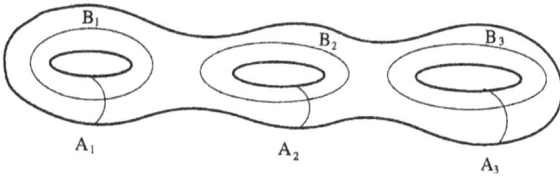

What gives Riemann surfaces their punch are the Θ-functions [53–56]; which define an n^g-dimensional Hilbert space consisting of independent meromorphic functions:

$$\Theta(A|T_{ij}) \equiv \sum_{n \in Z^g} \exp\left[i\pi\left(\sum_{ij} n_i T_{ij} n_j + 2\pi i \sum_j A_j n_j\right)\right], \quad \text{(C.2)}$$

where $A_j \equiv \int_{P_0}^{P} d\omega_j$ and the $T_{ij} \equiv \oint d\omega_j(B_i)$ are the "periods" for the Riemann surface obtained by integrating one of the g algebraically independent rational differentials on the Riemann surface around one of its "B" cycles. These functions are not L-periodic, but L-automorphic:

$$\Theta\left(A + \sum_j T_{ij} m_j, |T_{ij}\right) = [\exp\left[i\pi\left(T_{ii} + 2\pi i A_i\right]^n \Theta(A|T_{ij}).$$

$$\text{(C.3)}$$

One thing that is remarkable about Θ-functions is that they define an embedding of a Riemann surface in a projective space. Because of the form of the prefactor in Eq. (C.3), the Θ-functions define an embedding of the complex torus $\mathbb{C}^g/L$ into a projective space of dimension n^g. The Θ-functions define an embedding of a Riemann surface into a projective space of dimension n^{2g}, and therefore can be considered as quantum wave functions on the Riemann surface. These functions are not periodic automorphic w.r.t to a complex lattice L; i.e. instead of being L-periodic, they get multiplied by a factor

$$e_\alpha(x) = \exp\left(\pi\left[H(x,\alpha) + \frac{1}{2}H(\alpha,\alpha)\right]\right), \qquad (C.4)$$

where $\alpha \in L$. The Hermitian form H is defined by the fact that its imaginary part is just the usual intersection form for homology cycles which maps L x L to the integers. The form (C.4) plays an important role in relating the theory of Riemann surfaces to the theory of algebraic varieties and quantum mechanics. This relations are encoded in two fundamental topology theorems due to Solomon Lefshetiz [54]. Of particular interest to us is that the exact multi-soliton solutions to the KdV and NLS equations can be expressed in terms of the Riemann $\Theta =$ function:

$$\theta(z) \equiv \sum_{\mu \in Z^g} \exp\left[2i\pi\left(\frac{1}{2}\mu T\mu + \mu z\right)\right], \qquad (C.5)$$

where $z \in \mathbb{C}^g$, $\mu \in \mathbb{Z}^g$, and $T \in \mathfrak{H}_s$ where $\mathfrak{H}_s$ is the "Siegel half-plane" consisting of symmetric complex g x g matrices with positive imaginary parts. The Θ-functions (C.5) are naturally associated with Riemann surfaces, which for our purposes will be the auxiliary Riemann surfaces associated with either the KdV equation or the NSE. The period matrix $T = \{\tau_{ij}\}$ is determined by the periods of abelian integrals around the g nontrivial "B type" homotopy cycles of a Riemann surface with genus g. (The periods around the "A type" cycles are conventionally normalized to 1.) The Θ-function is not periodic on $\mathbb{C}^g$ but automorphic:

$$\theta(z + \alpha) = e^{2\pi i\left[-\mu \cdot z - \frac{1}{2}\mu T\mu\right]}\theta(z), \qquad (C.6)$$

where $\alpha \in \mathbb{C}^g$ can always be in the form $\alpha = I\mu' + T\mu$, and $\mu, \mu' \in \mathbb{Z}^g$. (These equations are written in different ways in different books on Riemann surfaces, but I follow Farkas and Kra). The Riemann Θ-functions are single valued and holomorphic on the complex torus $\mathbb{C}^g/\Lambda$, where Λ is the $2g$ dimensional real lattice defined by the periods of abelian integrals of the allowed g independent rational differentials [55] on a Riemann surface. A remarkable property of Θ-functions is that although they are holomorphic functions on $\mathbb{C}^g/\Lambda$ the ratio of Θ-functions with different values of the displacement β in (6.40) can be used to define by pullback from $\mathbb{C}^g/\Lambda$ to a corresponding Riemann surface a rational function on a Riemann surface that provides the prefactor (and scattering data) for the solution of the Lax equation for the KdV equation and NSE. The exact expression for this prefactor in terms of Θ-functions can be found in the literature on the NSE equation, but has the general form

$$\phi(z) = \frac{\theta(z+\alpha)\theta(z-\alpha)}{\theta(z+\beta)\theta(z-\beta)}. \tag{C.7}$$

The function $\phi(z)$ is a periodic meromorphic function on $\mathbb{C}^g$ with $2g$ period vectors $\{e_1, \ldots, e_g; \tau_1, \ldots, \tau_g\}$ (see Farkas and Kra). Remarkably, the numerator and denominator of (C.8) also provide a map into a projective space P^N where $N = 2^g - 1$. The basis vectors for the image of this map are provided by the shifted Riemann Θ-functions [54,55] where the integers μ and μ' are defined modulo 2, i.e. $\mu, \mu' \in \mathbb{Z}^g/2\mathbb{Z}^g$. Defining $\mu = \varepsilon/2$, then the basis vectors within P^N are called first-order Θ-functions with integer characteristics $[\varepsilon, \varepsilon']$ [54–57]:

$$\theta \begin{bmatrix} \epsilon \\ \epsilon' \end{bmatrix} (z|T) = \exp\left\{ \frac{\pi i}{2} \left[\frac{1}{4}\epsilon T\epsilon + \epsilon z + \frac{1}{2}\epsilon\epsilon' \right] \right\} \theta(z). \tag{C.8}$$

As the integer characteristics $[\varepsilon, \varepsilon']$ run over the coset labels (0,1) for $\mathbb{Z}^g/2\mathbb{Z}^g$ the "theta-null-werte", i.e. the values of $\theta \begin{bmatrix} \epsilon \\ \epsilon' \end{bmatrix} (z|T)$ at $z = 0$, define the generators for a 2^{2g} dimensional representation of the Heisenberg group! In this representation, the cosets $\mathbb{Z}^g/2\mathbb{Z}^g$ play the role of t in (6.2). The action of the translation part of the

Weyl–Heisenberg group on the Θ-functions is

$$\theta \begin{bmatrix} \epsilon \\ \epsilon' \end{bmatrix} (z + T_k | T) = \exp \left\{ 2\pi i \left[-z_k - \frac{T_{kk}}{2} - \frac{\epsilon'_k}{2} \right] \right\} \theta \begin{bmatrix} \epsilon \\ \epsilon' \end{bmatrix} (z | T)$$

(C.9)

This use of the explicit Weyl–Heisenberg generators allow one to calculate analytically the reward function for possibly all optimal control and RL of interest.

D. The Eightfold Way

As a precursor to our quantum approach to Bayesian inference, it was discovered [72] during the flurry of activity following the paper of Gardner *et al.* [73] on the inverse scattering solution of the KdV equation that the multi-soliton solutions of the KdV equation have a very pretty purely quantum interpretation in terms of the energy eigenstates for an array of quantum oscillators. This remarkable development is based on a method introduced by Murray Gell-Mann and Yuval Ne'eman [118] for constructing representations of SU(3) using 3 sets of fermion creation and annihilation operators. This method is based on the fact that representations for SU(N) can always be classified using N-1 SU(2) representations. In the case of SU(3), the two privileged SU(2) groups are traditionally referred to as "I-spin" and "U-spin". We are indebted to Gell–Mann for introducing the name "Eightfold Way" for the 8 generators of SU(3), which in addition to 3 "I-spin" and 3 "U-spin" operators employ 3 "We spin" operators, and in addition, the Buddha (channeled through James Joyce) conveyed to him the names up-quark, down-quark, and strange quark for the 3 kinds of fermionic operators needed for SU(3). Each SU(3) representation consists of a certain number of I-spin multiplets with well-defined numbers of each kind of quark. The total number of quarks minus anti-quarks attached to an SU(3) representation is known as the baryon number. A natural way to truncate the Hilbert space of a 3D oscillator is to fix the baryon number.

Ironically as a scheme for modeling nuclear particles as a loose assembly of quarks, the "Eightfold Way" turned out to not be of any fundamental importance for elementary particle physics. On the other hand, the use of the Gell–Mann–Ne'eman fermionic operators to describe the representations of SU(3) turns out to be of great importance for our program of translating Bayesian learning into quantum mechanics. The N fermion creation and annihilation operators that are used to construct representations for SU(N) satisfy anti-commutation relations:

$$\{\psi_m, \psi_n\} = \{\psi_m^*, \psi_n^*\} = 0, \{\psi_m, \psi_n^*\} = \delta_{nm}. \qquad (D.1)$$

Expressions of the form $\sum_{nm} c_{nm} \psi_m \psi_n^*$ then form a Lie algebra that acts in vector spaces $V = \sum c_n \psi_n$ and $V^* = \sum c_n \psi_n^*$ as

$$\{\psi_m \psi_n^*, \psi_p\} = \delta_{np} \psi_m, \{\psi_m \psi_n^*, \psi_p^*\} = -\delta_{mp} \psi_n^*. \qquad (D.2)$$

Exponentiation of this Clifford Lie algebra leads to continuous group $G(V, V^*)$:

$$\exp(t\psi_m \psi_n^*) = 1 + t\psi_m \psi_n^*. \qquad (D.3)$$

If we now fix a set of parameters $\{t_i\}$ (the "times" for the multiple action-angle flows associated with an integrable system) for a continuous Lie group, then one can define [154] as a fermionic analog of the Lie–Poisson Hamiltonian:

$$H(t) \equiv \sum_{l=1}^{\infty} t_l \sum_{n=-\infty}^{\infty} \psi_n \psi_{n+l}^*. \qquad (D.4)$$

If $g \in G(V, V^*)$, then

$$g(t) = e^{H(t)} g(0) e^{-H(t)} \qquad (D.5)$$

and the KdV τ-function introduced in Eq. (5.16) is [154]

$$\tau(x, g) = < g(x) > = < e^{H(x)} g(0) >, \qquad (D.6)$$

where the brackets refer to the ground state expectation value for an array of quantum oscillators. This τ-function is the glue that ties together the long-term behavior of an integrable system with local behavior represented by the Hamiltonian H. Of course, in practice the sums over l and n in the expression for $H(t)$ would have to be truncated in practice, not to mention the difficulties of representing fermions in a practical computational setting. Nevertheless, we have a setup which in principle would allow the τ and Baker functions for the KdV equation to be exactly evaluated using an array of quantum oscillators. Whether this is of any practical value remains to be seen. However, the "I-spin", "U-spin". and "We-spin" lines in the SU(3) root diagram used in the Eightfold Way make a ubiquitous appearance in the RH approach to solving nonlinear PDES. The reason for this is that that these axes play an important role [69] in defining the boundary separating the domains of the holomorphic functions which are used in Riemann–Hilbert approach to finding analytic solutions for integrable nonlinear PDEs.

E. Quantum Theory of Brownian Motion

E.1. *Quantum dynamics a la Feynman–Vernon Keldysh*

The theory of classical dynamics with noise was founded long before quantum mechanics by Fokker and Planck. In the 1950s, Wiener and Kac developed a classical path integral formulation of the Fokker–Planck theory of noise. In the 1960s, Feynman and Vernon [47] developed a quantum version. For an environment consisting of a classical noise source (e.g. a GP). The Feynman–Keldysh propagator J for the quantum mechanical density matrix $\sum_n \psi_n(x)\psi_n^*(x')e^{-En/kT}$ is

$$J = \iint e^{i\{S[x(t)]-S[x'(t)]\}/\hbar} \tag{E.1}$$

$$\times \exp\left\{-\int_0^T \int_0^t (x(t') - A(t,t')(x(t) - x'(t))dtdt'\right\} Dx(t)Dx'(t)$$

where $S[x(t)]$ is the classical action for the system and $A(t,t')$ is the autocorrelation function for the noise. If instead of a classical noise signal the quantum system is coupled to a quantum environment, then the real exponential in the formula for J is replaced by the complex valued influence functional $F\{[x(t), x'(t)]\}$ and $A(t,t')$ is replaced by a complex function $\alpha(t,t')$. The exponential factor in the density matrix propagator J can also be thought of as an overlap integral for final and initial states for the forward and backward 2nd oscillator array as a functional of $x(t)$ and $x'(t)$:

$$F[x(t)x'(t)] = \int \psi_Y(y_b)\psi_Y'^*(y_b)dY_b.$$

Another simple problem where the path integral focuses on the classical path is the harmonic oscillator. In this case the Feynman path integral [47] becomes

$$J(x_a, t_b; x_a, t_a) = \left(\frac{m\omega}{2\pi i\hbar \sin \omega t}\right)^{\frac{1}{2}}$$

$$\times \exp\left\{\frac{im\omega}{2\hbar\sin\omega t}\left[(x_a^2 + x_b^2)\cos\omega\, t - 2x_ax_b\right]\right\}.$$

$$\text{(E.2)}$$

As is the case for a free particle, the phase of the exponential is just the classical action measured in units of $\hbar$. This means that as a first approximation the quantum motion of an array of coupled oscillators can be calculated classically. This has the consequence that the quantum dynamics of an array of linearly coupled quantum oscillators can be efficiently simulated using conventional computers.

The Feynman–Keldysh prescription for the quantum dynamics of a system subject to influences by an "environment" is to replace the original Feynman path integral with a double path integral.

$$\int e^{iS[q(t)]}Dq(t) \rightarrow \iint e^{i\{S[q(t)]-S[q'\{t\}]\}}, F[q(t),q'(t)]Dq(t)Dq'(t),$$

$$\text{(E.3)}$$

where $F(q,q')$ represents the effect of the second quantum system; viz. the measuring apparatus. The exact form for $F(q,q')$ depends on the details of the second quantum system. However, in general one can write

$$F[x(t)x'(t)] = \exp\left\{-\int_0^T \int_0^t (x(t')\,\alpha(t,(t')) - x'(t')\alpha^*(t,t'))\right.$$

$$\left. - (x(t')-x'(t)\}dtdt'\right\},$$

where $\alpha(t,t')$ is a complex function that plays much the same role as the real autocorrelation function for the signal. In the case where the environment consists of a pure classical noise, usefulness of the

Feynman–Keldysh double path integral derives from the fact that it allows one to precisely describe the time evolution of the density matrix of a quantum system interacting with an environment. Green's function for describing the time evolution of the density matrix of a single harmonic oscillator is

$$J(x, y, x_0, x_0, t) = \iint \exp\left(\frac{im}{2\hbar}(\dot{x}^2 - \omega_0^2 x^2 - \dot{y}^2 + \omega_0^2 y^2)\right), \quad (E.4)$$

where ω_0 is the oscillator angular frequency. When this oscillator is coupled to an environment, Eq. (E.4) becomes

$$J(t_1, t_0) = \iint \exp\left\{-i\int_0^{t_1}\int_{t_0}^t \left\{\frac{M}{2\hbar}(\dot{q}^2 - \omega_0^2 q^2 - \dot{\bar{q}}^2 + \omega_0^2 \bar{q}^2)\right.\right.$$

$$- [\alpha(t, t')q(t') - \alpha^*(t, t')\bar{q}(t')][q(t) - \bar{q}(t)]dt\, dt'\Big\} \mathcal{D}q(t)\mathcal{D}\bar{q},$$

$$(E.5)$$

where $\alpha(t, t')$ is a complex function that will play much the same role as the real autocorrelation function $A(t, t')$ for a valued random time real signal. When the environment consists of another harmonic oscillator with level spacing Δ, the influence function is

$$F(x, y) \cong \exp -\frac{g^2}{2\Delta}\left(\frac{m\omega_0^2}{2\hbar}\right)$$

$$\times\left(\int_{t_0}^{t_0+T}\int_{t_0}^t \left[\left(e^{-i\Delta(t-t')}x(t') - e^{i\Delta(t-t')}y(t')\right)\right.\right.$$

$$\times(x(t) - y(t))\Big] dt'dt\Bigg)$$

(which in general implies non-Markovian quantum dynamics for the density matrix):

$$\frac{\partial\rho}{\partial t} = \left(\frac{i\hbar}{2L}\left(\frac{\partial^2\rho}{\partial Q^2} + \frac{\partial^2\rho}{\partial Q'^2}\right) + \left(\frac{iL\omega_0^2}{2\hbar}(Q^2 - Q'^2)\right.\right.$$

$$- \frac{C^2 \ln \Delta_{\max}/\Delta_{\min}}{\hbar^2 \Delta_{\max}}\left[Q(t) - Q'^{(t)}\right]\sum_{t_i=-\infty}^{t_i-t}$$

$$\times\int_{t_i-\tau}^{t_i}(Q(s) - Q'(s))ds\Bigg)\rho. \quad (E.6)$$

A general form for $F(q, q')$ that applies in many situations, and roughly speaking provides a quantum parallel for Gaussian noise, is

$$
F[q(t), q'(t)]
$$
$$
= \exp\left\{ -\int_0^T \int_0^t [\alpha(t, t')q(t') - \alpha^*(t, t')q'(t')][q(t) - q'(t)]dt'dt \right\},
$$
$$(E.7)$$

where $\alpha(t, t')$ is a complex function that plays much the same role in quantum mechanics as the autocorrelation function for Gaussian processes. The exact relation of $\alpha(t, t')$ to classical noise can be understood by looking at matrix elements of the quadratic functional of $q(t)$ and $q'(t)$ in the Hilbert space spanned by energy eigenstates of an array of quantum oscillators. For example,

$$
\int e^{S[q(t)] - S[q'(t)]} \left\{ \int_0^T \int_0^t \alpha(t, t')q(t')[q(t) - q'(t)]dt'dt \right\} Dq(t)Dq(t)
$$
$$
= -\int_0^T \int_0^t \alpha(t, t') < m|q(t)|n >< m|q(t')|n > dt'dt, \qquad (E.8)
$$

When the environment consists of quantum oscillators, $\alpha(t, t')$ has the form

$$
\alpha(t, t') = -\sum_{\omega_i} \frac{g_i^2}{\hbar^2} e^{-i\omega_i(t - t')}, \qquad (E.9)
$$

where the ω_is are the frequencies of the oscillators in the 2nd oscillator array making up the "environment". For a single harmonic oscillator and an environment consisting of oscillators with frequencies ω_i, where $\Delta_i = \hbar\omega_i$ is the level spacing, by analogy with the classical Wiener filter one might assume that $\alpha(t, t')$ has a piece that represents the signal and a piece that represents the noise. As a reminder the Wiener filter $H_F(t, s)$, described in Chapter 2, is obtained as a ratio of Laplace transforms of the signal correlation $K(s, t)$ and $R(s, t)$, the sum of the signal and noise correlation functions. If we add a classical noise term, $R(s,t)$ becomes

$$R(t,t') = -\sum_{\omega_i} \frac{g_i^2}{\hbar^2} e^{-i\omega_i|t-t'|} + A(t,t'), \qquad (E.10)$$

The kernel that appears in Wiener's integral expression for a signal accompanied by white noise is determined by the Wiener–Hopf factorization of the bilateral Laplace (not Fourier!) transform $R(s)$ of $R(t)$:

$$R(s) = \int_{-\infty}^{\infty} [\delta(t) + K(t)] \exp(-st)ds, \qquad (E.11)$$

where $K(t,x)$ is either a causal or anti-causal kernel:

$$K(t,s) = \sum_i \alpha_i \exp(-\beta_i t)\exp(\beta_i s), t \geq s,$$
$$K(t,s) = \sum_i \alpha_i \exp(-\beta_i s)\exp(\beta_i t), t \leq s. \qquad (E.12)$$

The Wiener–Hopf Eq. (A.6) allows one to extract features of the environment using the filter

$$\hat{Z}(t) = H_F(t,t')Y(t'), \qquad (E.13)$$

where $H_F(s)$ is a ratio of Laplace transforms of the $K(s,t)$ and the classical $R(s,t)$:

$$H_F(s) = \frac{R+(y)}{1 + R^+(y)}. \qquad (E.14)$$

E.2.   *Stochastic influence functions*

In general, the noise in a real quantum system is not Gaussian. For example, in superconducting quantum systems $1/f$ noise can be important. It is believed that this might be due to the coupling of the system to two level systems (TLSs). In case of a quantum oscillator, the Hamiltonian would be

$$H = \hbar\omega_r\left(a^\uparrow a + \frac{1}{2}\right) + \sum_j\left[\frac{1}{2}\hbar\Delta_j\sigma_j^z + ig\sigma_j^y(a^\uparrow - a)\right]$$

$$\dot{a} = i\omega_0 - i\sum_j\frac{g_i}{\hbar}\left(\sigma_j^+ e^{-i\Delta(t-t_0)} - \sigma_j^- e^{i\Delta(t-t_0)}\right) + F(t)$$

$$F(t) = -\sum_j\frac{g^2}{\hbar^2}\int_{t_0}^t\sin(t - t')\sigma_j^z(t')\left[a^+(t') - a(t')\right]dt'$$

Because the "starting" times t_0 for the episodes of coherent evolution for each TLS are randomly distributed, the influence function will be a product of influence functions for each TLS:

$$F(t) \approx \exp - \sum_j\frac{g^2 j}{\hbar^2}\left(\int_{t_{j0}}^t\int_{t_{j0}}^{t'}\left[e^{-i\Delta_j(t-t')}x(t') - e^{i\Delta_j(t-t')}y(t')\right]\right.$$

$$\left. \times\left[x(t) - y(t)\right]dt'dt\right).$$

Modeling the coupling of a TLs to a quantum oscillator as the $n = 0.1$ levels of a harmonic oscillator yields a density matrix evolution equation:

$$\frac{\partial\rho}{\partial t} = \left(\frac{i\hbar}{2L}\left(\frac{\partial^2\rho}{\partial Q^2} + \frac{\partial^2\rho}{\partial Q'^2}\right) + \frac{iL\omega_0^2}{2\hbar}(Q^2 - Q'^2) - \frac{g^2}{\hbar^2}\right.$$

$$\times \exp\left(-\frac{g^2}{\hbar^2}\left(\int_{t_0}^t Q(t')\int_{t_0}^{t'}Q(s)e^{i\Delta(t'-s)/\hbar}dsdt'\right.\right.$$

$$\left.\left.+ \int_{t_0}^t Q'(t')\int_{t_0}^{t'}Q'(s)e^{i\Delta(t'-s)\hbar}dsdt'\right)\right)$$

$$\left.\times\left(\int_{t_0}^t(Q(t)Q(s)e^{-i\Delta(t-s)/\hbar} + Q'(t)Q'(s)e^{i\Delta(t-s)/\hbar})ds\right)\right)\rho.$$

The influence functional can now be regarded [155] as a random function of time whose fluctuations can be measured by the autocorrelation function for oscillator amplitude.

References

[1] J. Hertz, A. Krogh, and R. Palmer, *Introduction to the Theory of Neural Computation* (Addison-Wesley, Boston, 1991).

[2] R. S. Sutton and A. G. Barto, *Reinforcement Learning* (MIT Press, 2018).

[3] K. J. Astrom and R. M. Murray, *Feedback Systems* (Princeton University Press, 2009)

[4] Pearl, *Probabilistic Reasoning in Intelligent Systems: Networks of Plausible Inference* (Morgan Kaufmann Publishers, San Francisco, 1998).

[5] B. Efron, *Large-Scale Inference: Empirical Bayes Methods for Estimation, Testing, and Prediction* (Cambridge University Press, 2010).

[6] D. Mackay, *Information Theory, Inference, and Machine, 18* (Cambridge University Press, New Delhi, 2005).

[7] C. M. Bishop, 'Model-based machine learning'. *Philosophical Transactions of the Royal Society* **A371** (2012), 2012022.

[8] P. Dayan, G. E. Hinton, R. Neal, and R. Zemel, 'The Helmholtz machine'. *Neural Computation* **7** (1995), 889.

[9] G. E. Hinton, P. Dayan, B. Frey, and R. Neal, 'The wake-sleep algorithm for unsupervised neural networks'. *Science* **7** (1995), 889.

[10] T. Yang and M. Shadlen, *Nature* **447** (2007), 1075.

[11] K. Pribram, *Brain and Perception* (Lawrence Erlbaum Associates, 1991).

[12] T. Kohonen, *Self-organizing Maps* (Springer, 1995).

[13] R. Ritter and K. Schulten, 'On the stationary state of Kohonen's self-organizing mapping'. *Biological Cybernetics* **54** (1986), 99.

[14] M. A. Nielsen and I. I. Chuang, *Quantum Computation and Quantum Information* (Cambridge University Press, 2000).

[15] L. D. Landau and I. M. Lifshitz, *Quantum Mechanics* (Pergamon Press, 1977).

[16] M. Nagasawa, *Schrodinger Equations and Diffusion Theory* (Birhauser, 1993).

[17] G. Chapline, 'Quantum mechanics and pattern recognition'. *International Journal of Quantum Information* **2** (2004), 295.

[18] Y. Aharonov, S. Popescu, and J. Tollaksenn, 'A time-symmetric formulation of quantum mechanics'. *Physics Today* (Nov. 2010).

[19] E. Todorov, 'General duality between optimal control and estimation'. *Proceeding of 47th IEEE Conference on Decision and Control* (IEEE, Cancun, Mexico, 2008).

[20] R. E. Bellman, *Dynamic Programming* (Princeton University Press, 1952).

[21] B. D. Anderson and J. B. Moore, *Optimal Control* (Prentice Hall, 1990).

[22] J. S. Dugdale, *Entropy and its Physical Meaning* (Taylor & Francis, 1996).

[23] C. Kittel, *Elementary Statistical Physics* (John Wiley & Sons, 1958).

[24] D. Mumford, 'Neuronal architectures for pattern-theoretic problems'. In *Large Scale Theories of the Cortex*, C. Koch and J. Davies (eds.) (MIT Press, 1994).

[25] G. Chaitin, *Algorithmic Information Theory* (Cambridge University Press, 1987).

[26] J. Rissanen, *Stochastic Complexity in Statistical Inquiry* (World Scientific, 1989).

[27] S. Kullback, *Information Theory and Statistics* (Wiley, 1959).

[28] B. Jedynak, P. Frazier, and R. Sznitman, 'Twenty questions with noise: Bayes optimal policies for entropy loss', *Journal of Application Probability* **49** (2011), 114.

[29] A. P. Dempster, N. M. Laird, and D. B. Rubin, 'Maximum likelihood from incomplete data via the EM algorithm'. *Proceeding of the Royal Statistical Society* **B39** (1977), 1.

[30] E. Todorov, 'Efficient computation of optimal actions'. *PNAS* **106** (2009), 11478.

[31] H. J. Kappen, 'Path Integrals and symmetry breaking for optimal control'. *Journal of Statistical Mechanics* **P11011** (2005).

[32] L. M. Laurie, *Feynman's Thesis: A New Approach to Quantum Mechanics* (World Scientific, 2005).

[33] W. Pauli, *Pauli Lecture on Physics* vol. 4 (MIT Press, 1978).

[34] J. H. Rose, 'Single-sided focusing of the time dependent Schrodinger equation'. *Physical Review A* **65** (2001), 127707.

[35] P. A. M Dirac, *The Principles of Quantum Mechanics* (Oxford University Press, 1958).

[36] G. Chapline, 'Quantum mechanics as self-organized information fusion'. *Philosophical Magazine* **B81** (2001), 541.

[37] M. Nauenberg, 'Quantum wave packets on Kepler elliptic orbits'. *Physical Review A* **40** (1989), 1133.

[38] R. Tyson, *Principles of Adaptive Optics* (Academic Press, 1991).

[39] H. J. Briegel, D. E. Brown, W. Dur, R. Raussendorf, and M. Nest, 'Measurement based quantum computation'. *Nature Physics* **5** (2009), 19.

[40] M. Planck, *The Theory of Heat Radiation* (Dover Publications, 1991).

[41] D. ter Haar, *The Old Quantum Theory* (Pergamon Press, 1967).

[42] B. L. Van Der Waerden, *Sources of Quantum Mechanics* (North-Holland Publishing, 1967).

[43] W. Heisenberg, *The Physical Principles of the Quantum Theory* (University of Chicago Press, 1930).

[44] H. Weyl, *Theory of Groups and Quantum Mechanics* (Martino Publishing, 2014).

[45] N. D. Mermin, 'Limits to quantum mechanics as a source of magic tricks'. *Physical Review Letters* **74** (1995), 831.

[46] M. Born, 'The statistical interpretation of quantum mechanics'. In *Nobel Lectures: Physics 1942–1962* (Elsevier Publishing, 1964).

[47] R. P. Feynman and A. Hibbs, *Quantum Mechanics and Path Integrals* (McGraw-Hill, 1965).

[48] T. A. Brun, 'A simple model of quantum trajectories'. arXiv.quat-ph/ 0108312v1 (2001).

[49] J. Schwinger, 'Quantum theory of Brownian Motion'. *Journal of Mathematical Physics* **2** (1961), 407.

[50] A. O. Caldeira and A. Leggett, 'Influence of damping on quantum coherence'. *Phyical Review* **46** (1985), 211.

[51] L. V. Keldysh, 'Diagram technique for nonequilibrium processes'. *Journal of Experimental and Theoretical Physics — Soviet Physics* **20** (1965), 1018.

[52] G. Chapline, 'Machine learning and quantum mechanics'. In *Advances in the Computational Sciences*, ed., E. Schwegler, B. Rubenstein, and S. Libby (eds.) (World Scientific, 2007).

[53] L. S. Shapley, 'Stochastic games'. *Mathematics* **39** (1953), 1095.

[54] P. Griffiths and J. Harris, *Principles of Algebraic Geometry* (John Wiley & Sons, 1978).

[55] D. Mumford, *Curves and their Jacobians* (University of Michigan Press, 1975).

[56] D. Mumford, *Tata Lectures on Theta* (Birkhauser, 1983).

[57] H. M. Farkas and I. Kra, *Riemann Surfaces* (Springer-Verlag, 1992).

[58] F. J. Dyson, 'Photon noise and atmospheric noise in active optical systems'. *Journal of the Optical Society of America A* **65** (1975), 551.

[59] T. Tao, 'Tribute to Israel Gelfand' (wordpress.com/2009/10/07/ Israel).

[60] V. A. Marčenko, 'The construction of the potential energy from the phases of the scattered waves'. *Mathematical Reviews* **17** (1955). (Publications, 1992).

[61] B. Noble, *Methods Based on the Wiener-Hopf Technique* (Chelsea Publishing Company, 1958). (wordpress.com/2009/10/07/Israel).

[62] J. Pearl, 'Theoretical impediments to Machine Learning with sparks from the causal revolution'. arXiv:1801.04016 (2018).

[63] B. Schoellkopf, 'Causality for Machine Learning'. arXiv:1911.10500 (2019).

[64] R. P. Feynman, *Quantum Electrodynamics* (Benjamin, 1961).

[65] S. Mandelstam, 'Introduction to string models and vertex functions'. In *Vertex Operators in Mathematics and Physics,* J. Lepowsky, S. Mandelstam, and I. M. Singer (eds.) (Springer-Verlag, 1985).

[66] G. Chapline, 'The bootstrap principle and equal-time commutators'. *Il Nuovo Cimento* **58** (1968), 1.

[67] N. I. Muskhelishvili, *Singular Integral Equations* (Dover, 1992).

[68] A. Its, 'The Riemann-Hilbert problem and integrable systems'. *American Mathematical Society Notices* **50** (2003), 1389.

[69] T. Trogdon and S. Olver, *Riemann-Hilbert Problems, Their Numerical Solution, and the Computation of Nonlinear Special Functions* (SIAM, 2016).

[70] A. K. Ghatak, R. L. Gallawa, and I. C. Goyal, Modified Airy Functions and WKB Solutions to the Wave Equation (US Government Printing Office, 1991).

[71] A. B. Migdal and V. P. Krainov, *Approximation Methods in Quantum Mechanics* (W. A. Benjamin, 1969).

[72] L. A. Dikey, *Soliton Equations and Hamiltonian Systems* (World Scientific, 1991).

[73] N. J. Hitchin, G. B. Segal, and R. S. Ward, *Integrable Systems* (Clarendon Press, 1999).

[74] P. Lax, 'Integrals of nonlinear equations of evolution and solitary waves'. In *Communications on Pure and Applied Mathematics* **21** (1968), 467.

[75] A. C. Newell, *Solitons in Mathematics and Physics* (Society for Industrial and Applied Mathematics, 1985).

[76] R. Hiroto, 'Exact solution for the Korteweg-deVries equation for multiple solitons'. *Physical Review Letters* **27** (1971), 1192.

[77] J. L. Burchnall and T. W. Chaundry, 'Commutative ordinary differential operators'. *Proceeding of Royal Society London Series* **A** **118** (1928), 557; H. F. Baker, 'Note on the paper by Burchnall and Chaundry'. *Ibid.* 584.

[78] M. Schneider, 'Bayesian linking of geosynchronous orbital debris tracks as seen by the LSST.' *Advances in Space Research* **49** (2012), 655.

[79] S. Godsil, 'The relationship between Markov Chain Monte Carlo methods for model uncertainty'. *Computational and Graphical Statistics* **10** (2001), 1.

[80] A. T. White, *Graphs Groups, and Surfaces* (North-Holland, 1984).

[81] N. Wiener, *Extrapolation, Interpolation, and Smoothing of Stationary Time Series, With Engineering Applications* (Technology Press and Wiley, 1949).

[82] S. M. Stigler, *The History of Statistics* (Harvard University Press, 1986).

[83] T. Kailath, 'A view of three decades of linear filtering theory', *IEEE Transactions on Information Theory* **IT-20** (1974), 146.

[84] D. Simon, *Optimal State Estimation* (Wiley, 2006).

[85] D. C. Lindberg, *Science in the Middle Ages* (University of Chicago Press, 1978).

[86] H. J. Sussmann and J. C. Williams, '300 years of optimal control from the Brachistochrone to the maximum principle'. *IEEE Control Systems* **27** (1997), 32.

[87] H. Schattler and U. Ledzewicz, *Geometric Control Theory* (Springer, 2012).

[88] L. Landau and E. M. Lifshitz, *Mechanics* (Pergamon Press, 1977).

[89] Z. Ma and C. W. Rowly, 'Lie-Poisson integrators: A Hamiltonian, variational approach'. *International Journal for Numerical Methods in Engineering* (Wiley, 2009).

[90] J. von Neumann and O. Morgenstern, *Theory of Games and Economic Behavior* (Princeton University Press, 1944).

[91] M. Gutzwiller, *Chaos in Classical and Quantum Mechanics* (Springer, 1990).

[92] I. Stewart, *Galois Theory* (Chapman and Hall, 1973).

[93] M. Kuga, *Galois' Dream* (Birkauser, 1993).

[94] D. Babbitt, 'Certain Hilbert spaces of analytic functions associated with the Heisenberg group'. In *Studies in Mathematical Physics*, E. Lieb, B. Simon, and A. S. Wightman (eds.) (Princeton University Press, 1971).

[95] V. Bargmann, 'On a Hilbert space of analytic functions and an associated integral transform'. *Communications on Pure and Applied Mathematics* **14** (1961), 199.

[96] S. Thangavelu, *Harmonic Analysis on the Heisenberg Group* (Birkhauser, 1998).

[97] N. Wiener, *Nonlinear Problems in Random Theory* (MIT Press, 1958).

[98] M. Schuld and N. Kiloran, 'Quantum Machine Learning models are Kernel methods'. *Physical Review Letters* **122**(4) (2019), 040504137.

[99] M. Schuld and F. Petruccione, *Machine Learning with Quantum Computers* (Springer, 2021).

[100] J. Polchinski, *String Theory* (Cambridge University Press, 1998).

[101] J. von Neumann, *Mathematical Foundations of Quantum Mechanics* (Princeton University Press, 1952).

[102] C. W. Helstrom, *Quantum Detection and Estimation Theory* (Academic Press, 1976).

[103] C. N. Yang, 'Concept of off-diagonal long-range order'. *Reviews of Modern Physics* **34** (1962), 694.

[104] G. Chapline, 'Theory of the superfluid transition in liquid helium'. *Physical Review* **A3** (1971), 1671.

[105] L. C. Thomas, *Games, Theory and Applications* (Dover, 2011).

[106] D. Silver *et al.*, 'General reinforcement algorithm that masters chess, shogi, and Go'. *Science* **362** (2018), 1140.

[107] T. Kohonen, 'Physiological interpretation of the self-organizing map algorithm'. *Neural Networks* **6** (1993), 895.

[108] G. Chapline, 'Spontaneous origin of topological complexity in self-organizing neural networks', *Network: Computational Neural Systems* **8** (1987), 185.

[109] R. Linsker, 'Self-organization in a perceptual network'. *Computer* **21** (1988), 105.

[110] G. Chapline, 'Minimum energy information fusion in sensor networks'. In *Proceeding of the 2nd International Conference on Information Fusion* (Society for Information Fusion, 1999).

[111] J. Schwinger, 'The Majorana formula'. In *Festshrift for I. I. Rabi* (Transactions on New York Academy of Sciences), **38** (1977).

[112] L. C. Biedenharn and J. D. Louck, *The Racah-Wigner Algebra in Quantum Theory* in Encyclopedia of Mathematics and its Applications, ed. G. Rota (Addison-Wesley, 1981).

[113] S. L. Braunstein and H. J. Kimble, 'Teleportation of continuous quantum variables'. *Physical Review Letters* **80** (1998), 869.

[114] E. P. Wigner, 'The unreasonable effectiveness of mathematics in the natural sciences'. *Communications Pure and Applied Mathematics* **13** (Feb. 1960), 1.

[115] G. Chapline, 'Is theoretical physics the same thing as mathematics?' *Looking Forward: Frontiers in Theoretical Science, Physics Reports* **315** (1999), 95.

[116] P. M. Morse and H. Feshbach, *Methods of Theoretical Physics* (McGraw Hill, 1953).

[117] R. Newton, *Inverse Schrodinger Scattering in Three Dimensions* (Springer-Verlag, New York, 1989).

[118] H. J. Lipkin, *Lie Groups for Pedestrians* (North-Holland, 1965).

[119] V. Barbu and S. S. Sritharan, 'H_∞ control theory of fluid dynamics'. *Proceedings of the Royal Society London A* **454** (1998), 3009.

[120] M. Thieullen and A. Vigot, 'Stochastic representation of the tau function with an application to the Korteweg-De Vries equation'. *Communications On Stochastic Analysis* **12** (2018), 1.

[121] R. Durbin and D. Willshaw, 'An analog approach to the traveling salesman problem'. *Nature* **689** (1987), 326.

[122] H. Kleinert, *Path Integrals* (World Scientific, 1995).

[123] B. O. Koopman, *Search and Screening: General Principles and Historical Applications* (Pergamon Press, 1980).

[124] D. C. Woods and S. M. Lewis, 'Design of experiments for screening'. arXiV:1510.05248 [stat.ME] (2015).

[125] E. Finley-Fruendlich, *Celestial Mechanics* (Pergamon Press, 1958).

[126] V. Korepin, N. M. Bogoliubov, and A. Izergin, *Quantum Inverse Scattering Method and Correlation Functions* (Cambridge University Press, 1993).

[127] H. Bethe, *Intermediate Quantum Mechanics* (W. A. Benjamin, 1964).

[128] H. Volklein, *Groups as Galois Groups* (Cambridge University Press, 1996).

[129] G. A. Jones, 'Bipartite graph embeddings, 'Riemann surfaces and Galois groups'. *Discrete Mathematics* **338** (2015), 1801.

[130] D. F. Walls and G. J. Milbum, *Quantum Optics* (Springer-Verlag, 1995).

[131] B. Friedrich and D. Herschbach, 'Stern and Gerlach: How a Bad Cigar helped reorient atomic physics'. *Physics Today* (Dec. 2003) 53.

[132] R. Raussendorf and H. Briegel, 'A one-way quantum computer'. *Physical Review Letters* **86** (2001), 5188.

[133] C. F. Barenghi and N. G. Parker, *A Primer on Quantum Fluids*, arXiv:1605580v2 [cond-mat.quant-gas] (2016).

[134] D. Bohm, 'A suggested interpretation of the quantum theory in terms of "Hidden Variables"'. *Physical Review* **85** (1952), 180.

[135] B. R. Frieden, 'Fisher information as the basis for the Schrodinger equation'. *American Journal of Physics* **57** (1989), 1004.

[136] M. Reginatto, 'Derivation of the equations of nonrelativistic quantum mechanics using the principle of minimum Fisher information'. *Physical Review* **A58** (1998), 1775.

[137] I. M. Khalatnikov, *Introduction to the Theory of Superfluidity* (W. A. Benjamin, 1965).

[138] G. Chapline, E. Hohfield, R. B. Laughlin, and D. Santiago, 'Quantum phase transitions and the breakdown of classical general relativity'. *Philosophical Magazine* **B81** (2001), 235.

[139] D. Christodoulou, 'Reversible and irreversible transformations in black-hole physics'. *Physical Review Letters* **25** (1970), 1596.

[140] F. D. M. Haldane, 'Model for a quantum Hall Effect without Landau levels'. *Physical Review Letters* **61** (2015), 1988.

[141] B. A. Bernevig and T. Hughes, *Topological Insulators and Topological Superconductors* (Princeton University Press, 2013).

[142] R. Jackiw and S-Y. Pi, 'Soliton solutions to the gauged nonlinear Schrodinger equation on a plane'. *Physical Review Letters* **64** (1990), 2969.

[143] G. Chapline and J. Dubois, 'Topological quantum image analysis'. *Proceeding of SPIE* **7342** (2009), 73420C.

[144] R. J. Baxter, *Exactly Solvable Models in Statistical Mechanics* (Academic Press, 1982).

[145] L. Kadanoff and A. C. Brown, 'Correlation functions on the critical lines of the Baxter and Ashkin-Teller Models'. *Annuals Physics* **121** (1979), 318.

[146] S-Y. Chu, 'Statistical origin of classical mechanics and quantum mechanics'. *Physical Review Letters* **71** (1993), 2847.1059.

[147] N. Ikeda and S. Taniguchi, 'Quadratic Wiener functionals, Kalman-Bucy filters, and the KdV equation'. *Advanced Studies in Pure Mathematics* **41** (2004), 167.

[148] K. Obermayer, H. Ritter, and K. Schulten, 'Large-scale simulations of self-organizing neural networks on parallel computers: Applications to biological modeling'. *Parallel Computing* **14** (1990), 381.

[149] P. Suppes, Z. Lu, and B. Han, 'Brain wave recognition of words'. *PNAS* **94** (1997), 14965.

[150] G. Chapline, C-Y. Fu, and S. Nagarajan, 'Inverse scattering approach to improving human pattern recognition'. *Proceeding of SPIE* (2000).

[151] C. E. Rasmussen and C. K. I. Williams, *Gaussian Processes for Machine Learning* (MIT Press, 2006).

[152] J. Ko and D. Fox, 'GP -Bayes filters: Bayesian filtering using Gaussian process prediction and observation models'. *Autonomous Robots* **27** (2009), 75.

[153] J. Lee *et al.*, 'Deep neural networks as gaussian processes' (*International Conference on Learning Representations* 2018, arXiV: 1711.00165v1 [stat.ML]).

[154] S. Frautschi, *Regge Poles and S-Matrix Theory* (W. A. Benjamin, 1963).

[155] G. Chapline and M. Otten, 'Bayesian searches and quantum oscillators'. *Proceedings of 3rd Workshop on Microwave Cavities and Detectors, April 2018* (Springer, 2020).

[156] J. Barrett and I. Naish-Guzman, 'The Ponzano-Regge model'. *Classical and Quantum Gravity* **26** (2009), 155014.

Index